AF589392

Hydrogeochemical and Statistical Behaviour of Hill Springs

THE AUTHORS

Dr. Hemant Kumar Joshi is presently working as Assistant Professor in Nanhi Pari Seemant Engineering Institute, Pithoragarh (A constituent institute of UTU, Dehradun, Uttarakhand). He has teaching experience of more than five years. He has contributed seven research papers in national and international journals of repute.

Dr. Narendra Singh Bhandari is currently Professor of Inorganic Chemistry in the Department of Chemistry, Kumaun University, S.S.J. Campus, Almora, Uttarakhand. He has a teaching and research experience of twenty nine years and his over forty research papers have been published in journals of national and international repute. He has contributed 3 chapters in different books. Presently, he is working in the field of adsorptive removal of heavy metal ions from synthetic waste water by using bio-sorbents prepared from waste plant biomass available in Kumaun hills.

Hydrogeochemical and Statistical Behaviour of Hill Springs

Dr. Hemant Kumar Joshi
Dr. Narendra Singh Bhandari

2018
Scholars World
A Division of
Astral International Pvt. Ltd.
New Delhi – 110 002

ISBN 9789387057654 (International Edition)

Published by : **Scholars World**
A Division of
Astral International Pvt. Ltd.
– ISO 9001:2015 Certified Company –
4736/23, Ansari Road, Darya Ganj
New Delhi-110 002
Ph. 011-4354 9197, 2327 8134
E-mail: info@astralint.com
Website: www.astralint.com

DEDICATED

TO

OUR PARENTS

Acknowledgements

We are grateful to Prof. Ganga Bisht, Head of the Chemistry, Prof. Lata Joshi, Dean, Faculty of Science, Kumaun University and faculty members, whose patience and kindness have been invaluable to us. We appreciate all their contributions of time, guidance, support, fruitful suggestions, constructive criticism, and innovative ideas in preparation of this manuscript.

We are forever indebted to our parents and family, for their constant source of support, emotional, moral, endlesspatience and encouragement, and this book would certainly not have existed without them.

One of the authors expresses his special thanks and gratitude to Council of Scientific and Industrial Research (CSIR), New Delhi for the financial support as Junior Research Fellowship (JRF). We acknowledge the contribution of Mr. Sanjay Kumar, Senior Technical Assistant, Vivekanand Parvtiy Krishi Anushandhan Sansthan, Almora, who provided valuable suggestions towards the successful completion of this work. We extend our thanks to Mr. Godhan Negi for his help during the field work.

Finally, we would like to thank all our friends for their continuous love and encouragement when it was most required.

Dr. Hemant Kumar Joshi

Dr. Narendra Singh Bhandari

Foreword

Being essential for the survival of life as we know it, water is among the most inestimable gift of nature. Through its glacial and non-glacial rivers, Uttrakhand is a major source of water fo the Gangetic plain, which support a hife population. Till now we have tended to take the availabiliy and quality of warer for granted, but recent time ha shown is that our reckless ise of this invaluable matiral resources has taken a heavy toll on both. Thus, we are increasongly realising that a rapidly escalating water crisis is upon us.

The two major components of the water crisis, namely its availability and quality, need to be handled separately. Taking care of the rapidly degrading quality of ground water needs to start with ascertaining its present quality, which is what this work focues on. I applaud the authors for selecting such a crucial issue and working painstakingly on it, generating primary data and presenting it very licidly. I hope the book motivates other researchers and academicians to select similar critical issues.

(Prof. Durgesh Pant)
Director
Uttarakhand Science Education and
Research Centre (USERC)
Govt. of Uttarakhand, Dehradun

Preface

Spring is any natural situation where water flows from an aquifer to the Earth's surface and is a component of hydrosphere. In hill areas, across the world ground water is available only in the form of natural springs, which is used by the people to meet their requirements. In Kumaun hills, where about 30-40% of the total population still dependent on natural springs for their daily water needs the importance of maintaining and monitoring spring water quality can not be ignored. So the evaluation of spring water quality is great significance in determining the suitability of natural spring water uses like dinking and irrigation. The present work is an outcome of research work conducted on spring water during pre-monsoon and post monsoon seasons in three consecutive years (2007-2009). It covers a detail account of hydro-geochemistry of hill springs along with their chemical groupings which will be meaningful addition of the existing knowledge on natural springs of the Kumaun region of Uttarakhand. The findings of the study will certainly give an updated feature regarding the springs and provide the scientific information, beneficial for the improvement of the quality as well as quantity of the spring water. In addition to this, this book will also provide information about the impact of the anthropogenic activities on the hydrochemistry of spring water.

This book will help in creating awareness among policy makers, water resource planner and users about water resources, hill springs in particular. A brief review of work reported on various aspect of chemical composition of spring/ground water, hydro-geochemical characterization, potabilty, irrigation and statistical studies is included in chapter-1, chapter-2 presents the various materials analytical methods employed in the study. Description of study area with location of springs is given in chapter-3. Hydro-geochemical characterization and classification of hill springs are the part of the chapter-4. Evaluation of springs for potability and irrigation is covers in the chapter-5. The statistical behavior of chemical parameters of springs is discussed in the chapter-6. The conclusion of the present work makes the chpater-7. At the end of each chapter, references concerned are mentioned.

The present book will be useful for students, researchers, academicians, policy makers and water resource planners.

Dr. Hemant K. Joshi

Dr. N.S. Bhandari

Contents

1

Introduction

Literature on the hydrogeochemical aspect of ground water (springs/hand pumps) is compiled in the introduction. Chapter one starts with early believes about water as an elements and its anomalous behavior. The concept of water contamination/pollution, their possible sources of water resources and chemical composition of ground water are also the part of this chapter. The section recent studies incorporates various reports on early and recent studies on physico-chemical parameters, hydro-geochemical analysis, potability, irrigation water quality and statistical investigations. The objectives of the present work, complete the chapter one.

Water is one of the most precious gifts and first creation of Almighty. Water is the first and foremost important constituents of each and every living being. The dawn of civilization took place near the water resources. Water covers almost two third of the earth's surface. Water is without doubt the most abundant, most accessible, and the most studied of all chemical compounds. Its omnipresence, it's crucially important for man's survival and its ability to transform so readily from the liquid to the solid and gaseous states has ensured its prominence in man's thinking from the earliest times. In a sense, water is a weird compound of nature.

1.1 Water not element but a compound

Aristotle considered water to be one of the four elements, alongside earth, air and fire and this believe in the fundamental and elementary nature of water persisted until the second half of the eighteenth century. In the end of the 18th century, during the age of relief, scientist started to research all things they did not know about water.

The epoch-making experiments of Henry Cavendish *et al.* in the year 1781 about the composition of water revealed that it is a compound of oxygen and hydrogen. In the year 1783, Antoine Lavoisier et al. succeeded in splitting water into hydrogen and oxygen and later in the year 1789 J. R. Deiman and A. Peats Van Troostwijk obtained the same results.

1.1.1 Water, the unique chemical

The properties of water make it unique in comparison to other compounds and some of its significant anomalous properties have been discussed by Das. The unique physical and chemical properties of water have allowed life to evolve in it. The following quote from Szent-Gyorgyi illustrates this point of view:

"That water functions in a variety of ways within a cell cannot be disputed. Life originated in water, is thriving in water, water being its' solvent and medium. It is the matrix of life."

Water is a universal solvent. It acts as a transporting medium for dissolved material into, through and out of the body of an organism. It is also an agent for the chemical transformation of the materials within the body. Plants obtain mineral nutrients from the soil, which are dissolved in water. The animals take their food in solid form, but before it is absorbed in the blood, it must be dissolved. Hence, for digestion, the water is an essential component and also a constituent in the circulatory, excretory and reproductive fluids of animals. Water also regulates the temperature of the living organisms.

Celebrated dictum of Thales of Miletus, a first known Greek philosopher, describes water as:

"It is water that in taking different forms constitutes the earth, the atmosphere, the sky, the mountains, the Gods and men, the beasts and the birds, the grass and the trees, animals down to worms, flies and ants. All these are but different forms of water. Meditate on water!"

1.1.2 Quality of water

Most of the water in the hydrosphere is in the very salty oceans, and almost all of the remainder is tied up in ice. That leaves relatively little surface or subsurface water as potential fresh water resources for biotic community. The quality of groundwater is expressed in terms of certain defined parameters and by the concentration of toxic elements or compounds whose presence may cause a health hazards to humans, domestic animals, and wild life. For human consumption as well as for commercial use, the quality of water is more important than its quantity. Water is called of good quality only when it is free from salinity, plants and animals waste, bacterial and chemical contamination.

1.2 Concept of contamination and pollution

Contaminants are something which causes contamination such as waste materials emptied into the source of water supply or noxious fumes into the air and thus 'to contaminate' means to render impure by contact or mixture.

Pollute is to make foul or unclean and pollutant that pollutes especially chemical or refuse material released into the atmosphere or water. Pollution simply is the accumulation of chemical compounds or anything else where it is not wanted.

1.2.1 Sources of contaminant/pollutant

Contaminants/pollutants are the substance or group of substances, which can be identified as the contamination/pollution causing. Main sources are:

- Substances actually toxic to human beings and fauna as lead, mercury, cadmium, and their salts including cyanides and various pesticides.
- Substances, which becomes toxic by biochemical transformation such as methylation of mercury or by bio concentration.
- The discharge of industrial effluents to river and septic tank soak way into permeable strata are also the point sources of contamination/pollution.
- Substances, which add to bottom deposits such as sewage effluent, waste water and thus increase the load of bio chemical oxygen demand of water. Discharge from small domestic sewage treatment plants, wash off of litter dust and dry fallout from urban areas to water resources may be the point sources of contamination/pollution.
- Substances having a detrimental effect on the physical appearance of water such as oil, detergent and foam etc. Domestic use of oil, detergent and foam may also be identified as point sources of contamination/pollution.
- Substances having a quality deteriorating effect at relatively high concentration in water such as minerals salts, water borne organism pathogenic to human beings.

There are other sources of pollution, which contaminate/pollute water but cannot be identified particularly. No particular person or corporate body can be held responsible for such type of pollution. So these sources are difficult to identify and control. Special research and administration action are needed for remedial and preventive measures to check such as sources of contamination/pollution of water.

1.3 Chemical composition of ground water

The chemical composition of ground water mainly relates to the minerals that react with it. Groundwater chemistry is largely a function of the mineral composition of the aquifer through which it flows. As groundwater moves along its path from recharge to discharge areas, a variety of hydro-geochemical processes alter its chemical composition. Reactions between groundwater and aquifer minerals have a significant role on water quality, which are also useful to understand the genesis of groundwater. As groundwater flows through sediments the process of cation exchange on clay layers causes decrease of calcium and/or magnesium and increase of sodium which resulted in gain or loss in total equivalents of cations in groundwater (equation-01):

$$\text{Ca, Mg Clay} + 2\,\text{Na}^{+} \xleftrightarrow[\text{Base Exchange}]{} 2\,\text{Na Clay} + \text{Ca}^{2+}, \text{Mg}^{2+} \qquad ...01$$

Ahmed and Ali have explained the process of mineral dissolution. The alkaline feature of groundwater causes dissolution of base compounds such as calcium bicarbonates. Dissolution of carbonate and sulfate minerals in the sediments result in high value of total hardness to ground water. High concentration of sodium has been associated to base exchange and leaching of sodium salts such as halite (NaCl) during the movement of groundwater through sediments. Higher values of sulfates in groundwater at the new reclaimed lands have been associated to dissolution of gypsum (equation-02) and potassium sulfates added to the soil as fertilizers.

$$CaSO_4.2H_2O \longrightarrow Ca^{2+} + SO_4^{2-} + 2H_2O \quad ....02$$

The source of calcium and magnesium in groundwater is dissolution of carbonate minerals such as calcite (equation-03), aragonite, and dolomite (equation-04) of the rocks.

$$CaCO_3 + H_2CO_3^- \longrightarrow 2HCO_3^- + Ca^{2+} \quad ...03$$

$$CaMg(CO_3)_2 + 2H_2CO_3 \longrightarrow Ca^{2+} + Mg^{2+} + 4HCO_3^- \quad ...04$$

The high TDS values have been attributed to leaching of salts from Plio-Pleistocene sediments containing salts of sulfates and chlorides.

The existing minerals for dissolution may also be gypsum ($CaSO_4.2H_2O$), anhydrite ($CaSO_4$), epsomite ($MgSO_4.7H_2O$), brucite ($Mg(OH)_2$), mirabilite ($Na_2SO_4.10H_2O$), halite (NaCl), and thenardite (Na_2SO_4).

The chemical analysis of groundwater reveals that the chemical composition varies between wide limits depending on the mineral composition of the regolith and underlying bedrock, and on anthropogenic activities involve in water system. The total chemical composition of groundwater consists of ions and molecules in the suspension; the chemical composition of the colloidal particles and dissolved gases whose concentration depends on the temperature of the water and their partial pressure in the atmosphere.

A recent reevaluation by Berner and Berner of the data compiled by Meybeck and Livingstone, indicates that the concentration of most major cations ions viz. Na^+, Ca^{2+}, Mg^{2+} and K^+ and cations *viz.* SO_4^{2-}, Cl^-, HCO_3^- and NO_3^- predominates into ground water which significantly increased by anthropogenic contamination.

1.4 Hydro-geochemical characterization

Potability is the main consideration of the groundwater. In many areas groundwater quality limits to supply of potable fresh water. Hence, to utilize and protect valuable water resources effectively and predict the change in groundwater environments, it is necessary to understand the hydro-geochemical characteristics of the groundwater and its evolution under natural water circulation processes.

In recent years, for many of the world towns or cities, groundwater seems to be the only source of fresh water to meet domestic, agricultural, and industrial needs. Earlier studies reported the importance of hydro-geochemical studies of groundwater in a particular region.

Jalali has reported the major hydro-geochemical processes that control the observed groundwater chemistry in the Alisadr, Hamadan, and western Iran.

Fianko and coworkers analyzed chemical constituents of Densu basin, Ghana, representing the country's greatest hydro structure with freshwater are found generally low in the North and high in the South. The order of relative abundance of major cations in the groundwater is $Na^+ > Ca^{2+} > Mg^{2+} > K^+$ while that of anions is $Cl^- > HCO_3^- > SO_4^{2-} > NO_3^-$. Four main chemical water types were delineated in the Basin. These include Ca–Mg–HCO_3, Mg–Ca–Cl, Na–Cl, and mixed waters in which neither a particular cation nor anion dominates. Silicate weathering and ion exchange are probably the main processes through which major ions enter the groundwater system. Anthropogenic activities were found to have greatly impacted negatively on the quality of the groundwater.

The hydro geochemical characterization of Coxilha das Lombas Aquifer, Brazil is reported by Herlinger and cowerkers. The concentration of most abundant ions usually followed the trend $Cl^- > HCO_3^- > SO_4^{2-} > Na^+$. According to Piper diagram, the samples were classified as alkali-sulfate-chloride-rich (73% of the samples) and calcium-bicarbonate-chloride-rich (37% of the samples). Hematite and goethite presented favorable conditions to precipitation.

Sadashivaiah and cowerkers studied Hydrochemical analysis and evaluation of groundwater quality in tumkur Taluk, Karnataka state. The water type that predominates in the study area is Ca-Mg-HCO_3 during both pre and post-monsoon seasons of the year 2006, based on hydro-chemical facies. This pattern is attributed to the geology of the area which comprises igneous rocks of crystalline nature, in which the major units were gneisses and granites. Ground water in the study area occurs under water table conditions in the weathered and fractured granite, Gneisses.

Hydro geochemical characteristics of shallow groundwater in a mid-western coastal aquifer system, Korea have been investigated by Jeen and cowerker. The chemical composition of the groundwater was characterized by high chloride concentration and high variations in cation concentrations due to the cation exchange reaction between aquifer minerals and seawater components. The groundwater is classified into Ca $(HCO_3)_2$, $CaCl_2$, and NaCl types based on its hydro geochemical characteristics. The groundwater from the alluvial aquifer shows higher salinity in the rainy season than the dry season while the groundwater in the bedrock aquifer shows lower salinity year-round.

Hydrochemical investigations in the Kalambaina Formation Sokoto Basin have been initiated by Alagbe and states that groundwater chemistry is predominantly of two facies. These are calcium–magnesium–bicarbonate (Ca–Mg–HCO_3) and calcium– magnesium–sulfate–chloride (Ca–Mg–SO_4–Cl) facies. The dissolution of calcium and magnesium carbonates as well as some human/land-use activities has been related to these facies.

Hydro geochemical investigations were carried out in Chithar River basin, Tamil Nadu, India by Subramani and cowerkers to identify the major geochemical processes that regulate groundwater composition and concluded that weathering of carbonate and silicate minerals and ion exchange reactions predominantly determines the major ion chemistry. Carbonate-rich rocks such as crystalline limestone, dolomitic limestone, calcgranulite and kankar (lime-rich weathered mantle overlies carbonate rocks) were the major sources for carbonate weathering.

Geochemical characterization of karst groundwater in the cradle of humankind world heritage site, South Africa has studied by Holland and cowerkers and found that water chemistry in the catchment is evolved from a Ca–Mg to Na + K cation predominance and from a HCO_3^- towards SO_4^{2-} or Cl^- anion predominance. The direction of these trends was consistent with increasing specific conductance and total dissolved solid content of the samples towards potential sources. The Ca–Mg–SO_4 facies samples were influenced by acid mine drainage and represent the highly mineralized water samples. The Ca–Mg–HCO_3 facies of pristine dolomitic water was not impacted by anthropogenic sources. According to themdominant water facies in the study area changed due to infiltration of polluted water from a Ca–Mg–HCO_3 type to an Mg–Ca–SO_4 or Na–SO_4 type. The geochemical processes responsible for the variation in groundwater quality were mixing, ion exchange and mineral dissolution. The results showed areas that have water chemistry due to natural water–rock interactions (pristine dolomitic water) while other areas were highly impacted by anthropogenic sources.

Hajalilou and coworker investigated hydro geochemical parameters and groundwater quality assessment in Marand Municipality, northwest of Iran. According to them Piper diagram and correlation matrix results indicated that the main type of groundwater was bicarbonate with influential role of alkaline earths. The correlation matrix between chemical variables indicated that there was the highest correlation between HCO_3^- and Mg^{2+} ions (more than 80%) in the spring samples. Hydrochemical facies of water samples mainly consist of Ca-Mg-HCO_3. Stiff diagrams also represent bicarbonate and calcium-magnesium ions as dominant ions of the groundwater. Hydrochemical facies of water samples mainly consist of Ca-Mg-HCO_3.

Al-Agha and coworkers evaluated the hydrochemical facies using the trilinear diagram for 200 water samples of Gaza Strip, Palestine. The Groundwater in the north and west was mostly characterized by Ca-Mg-HCO_3 facies (alkaline water), and in the south and east by Na-Cl-SO_4 facies (saline water) and sand dunes and rainfall were the major factors controlling the distribution of hydrochemical facies.

The Pleistocene wells and springs, Jericho area, Palestine was studied by Khayat and cowerkers. Springs in the west were characterized by Ca–Mg–CO_3 water with a low degree of mineralization (~548 mg TDS/1). These springs show low Cl contents with molar ratios of ~1 for Na/Cl, ~1.2 for Mg/Cl and ~0.18 for SO_4/Cl. The groundwater from wells in the west of Wadi Qilt shows intermediate values between the fresh and saline end members, with a water type of Mg–Na–Ca–Cl–HCO_3.

Ramkumar and cowerkers determined the groundwater quality in parts of Vedaraniyam region, Tamil Nadu, India in order to understand the hydrogeochemistry of the water and found that most of the well water falls in Na-Cl type. They have reported that some well water having higher concentration of pH due to weathering of plagioclase feldspar by dissolved atmospheric carbon dioxide that will release sodium and calcium which progressively increase the pH and alkalinity.

The hydro geochemical data of ground waters of the different aquifers of the Ras Sudr-Abu Zenima area, southwest Sinai, Egypt were examined by El-Fiky Anwar to determine the main factors controlling the groundwater chemistry and salinity as well as its hydro geochemical evolution. These samples were characterized by the dominance of Cl+SO_4 over HCO_3. Three samples showed dominance of Na over Ca+Mg. The rest of the samples showed dominance of Ca + Mg over Na. Durov diagram plot revealed that the groundwater has been evolved from Ca-HCO_3 recharge water through mixing with the pre-existing groundwater to give mixed water of Mg-SO_4 and Mg-Cl types that eventually reached a final stage of evolution represented by a Na-Cl water type.

Silva-Filho and coworker studied a geochemical characterization of the coastal aquifer from the oceanic region of Niterói, Rio de Janeiro, Southeast Brazil, using hydro geochemical, correlation matrices and factor analysis. Based on the hydrochemistry, the groundwater was classified into three types: Group 1 (53%) belongs to Na–Ca–Cl–HCO_3 facies, Group 2 (20%) belongs to Na–Ca–Cl–SO_4 facies and Group 3 (27%) belongs to an intermediate Na–Mg–Ca–HCO3–Cl–SO_4 facies. The main source of bicarbonate (alkalinity) for groundwater in this region was the decay of sediment/soil organic matter Na^+ and Ca^{2+} were dominant cations.

Petalas and cowerkers studied the characterization of the environmental hydrochemistry in the broader Sapes area–Thrace region. Certain characteristic chemical types are reported Ca-Mg-HCO_3-SO_4, Ca-Mg-SO_4-HCO_3, Ca-SO_4, Ca-Mg-SO_4, Ca-Na-Cl-HCO_3 and Na-Cl. Calcium was the dominant Cation. The order of dominance for the heavy metals in surface and ground waters follows the order: Fe>Mn>Zn>Ni>Cu.

Contamination of groundwater in Tirupur region, Tamil Nadu, India has been reported by Karuppapillai and coworkers in various places. The abundance of major ions in groundwater was found in the given order: Na^+>Ca^{2+}>K+>Mg^{2+}=HCO_3^- >Cl^->SO_4^{2-}>NO_3^-. The water reported for the water types are calcium, sodium, bicarbonate, chloride (Ca-Na-HCO_3-Cl); sodium, bicarbonate, chloride (Na-HCO_3 -Cl); sodium, chloride (Na-Cl); calcium, sodium, chloride (Ca-Na-Cl); calcium, bicarbonate, chloride (Ca-HCO_3-Cl); calcium, bicarbonate (Ca-HCO_3); calcium, sodium, bicarbonate (Ca-Na-HCO_3); calcium, chloride (Ca-Cl); and sodium, bicarbonate (Na-HCO_3).

An attempt was made by Laluraj and coworkers to study the groundwater chemistry of aquifers, which lie along the coastal zone of central Kerala. The trilinear diagram of this study was classified into four hydrochemical facies based on the dominance of different cations and anions: facies 1: Ca^{2+}-Mg^{2+}-HCO_3^- type I; facies 2: Na^+-K^+-Ca^{2+}-HCO_3^- type II; facies 3: Na^+-K^+-Cl^--SO_4^{2-} type III and facies 4: Ca^{2+}-Mg^{2+}-Cl^--SO_4^{2-} type IV.

The seasonal variations of the chemical budget of ions in the proximity of River Cooum, Chennai, India were determined from the hydrochemical investigation of the groundwater by Giridharan and cowerkers. Hydrochemical characteristics of ions in the groundwater were studied using 1:1 equiline diagrams. Nature of the water samples was determined using the Piper trilinear diagram. The groundwater during the pre monsoon shows the dominance of excessive evaporation, silicate weathering and anthropogenic activities whereas in post monsoon, dilution predominates over that of other factors.

The lower Varuna River basin in Varanasi district in the central Ganga plain is studied by Raju and coworkers which is a highly productive agricultural area, and is also one of the fast growing urban areas in India. Hydro-geochemical facies of ground waters and the dominant hydrochemical facies is Ca-Mg-HCO_3 with appreciable percentage of the water having mixed facies. In the groundwater chemistry, the order of cation abundance is Na>Ca>Mg>K except in some samples where Mg replaces Na and in anionic chemistry the order is HCO_3>Cl>NO_3>SO_4>F.

Groundwater qualities of coastal aquifers in the Chennam-Pallippuram Panchayath of Alappuzha district, Kerala have extensively been monitored by Manjusree and coworkers to assess its suitability in relation to domestic and agricultural uses. Hydro geochemical processes controlling the water chemistry were precipitation rather than rock-water interaction. The characteristics of the water samples varied from oligohaline to brackish. The abundance of other cations and anions are $Ca^{2+}>Mg^{2+}>Na^{+}>K^{+}$ and $HCO_3^{-}>Cl^{-}>SO_4^{2-}$.

Mohsen Jalali conducted a study to evaluate factors regulating groundwater quality in an area and to understand the sources of dissolved ions. The chemical compositions of the ground water are dominated by Na^+, Ca^{2+}, HCO_3^-, Cl^- and SO_4^{2-}, which have been derived largely from natural chemical weathering of carbonates, gypsum and anthropogenic activities of fertilizer's sources.

1.5 Concern for Water Quality

The consequences of the contamination of the environment reported by Rachel Carson in a series of articles in The New Yorker in 1960 have generated widespread public concern and discussion about the use of pesticides in agriculture. In 1962 she wrote Silent Spring, which for the first time mentioned the potential dangers of pesticides as contaminants of water resources. The Silent Spring created public awareness for environment and initiated environmental movements in the world.

1.5.1 Potability

Rao has studied the nitrate pollution and its distribution in the groundwater of Srikakulam District, Andhra Pradesh, India.

Fianko and coworkers have studied the various sources of contaminants as well as assessed the physical and chemical quality of the groundwater in Tema District, Ghana.

The physico-chemical studies on Sokoto Basin showed water of fairly good quality. Concentration of the total dissolved solids found was between 130 and 2,340 mg/l. NO_3^- is on the higher side of the permissible limits, indicating pollution in such areas.

The impact of pre- and post-monsoons on the groundwater quality of shallow coastal aquifers of Rameswaram Island in India is studied by T. Ramachandramoorthy. The water quality index advocated the transfer of groundwater quality from unacceptable status in the pre-monsoon into an acceptable status in the post-monsoon.

Evaluation of water quality index for drinking purposes for river Netravathi, Mangalore, South India, was done by Avvannavar and cowerkers. The water quality index (WQI), using six water quality parameters dissolved oxygen (DO), biochemical oxygen demand (BOD), most probable number (MPN), turbidity, total dissolved solids (TDS) and pH measured at eight different stations along the river basin. Rating curves were drawn based on the tolerance limits of inland waters and health point of view. Bhargava WQI method and harmonic mean WQI method were used to find out overall WQI along the stretch of the river basin.

Abdul Jameel has evaluated the drinking water quality in Truchirapalli, Tamilnadu. 24 spots were identified within the radius of about 20 km. The study revealed that the pH of water samples was mild alkaline in the range of 7.6 to 8.6. In some water samples, conductivity was more than the permissible limit (>1500 mho/cm) indicating the presence of dissolve ionic substances. All the ground water samples except that of Cauvery River were having amounts of suspended dissolved and total solids indicating high pollution. Nearly ten out of twenty stations had the hardness beyond the permissible limits for both total and permanent hardness. Six out of twenty ground water bodies had abnormally high value of nitrate-nitrogen exceeding 50 mg/l and among them three samples had greater than 100 mg/l, though sulfate concentration in all the water samples was within the permissible limits, but the chloride was high (>600 mg/l). Sodium and potassium were the common cations present in water. The maximum concentration of calcium and magnesium were found to be 453 mg/l and 166 mg/l respectively. DO was less than 5 mg/l. Since waters were unsuitable for drinking purposes methods to improve the water quality have been suggested.

Sudhir Kumar and cowerkers studied distribution of nitrate in the drinking water samples of various parts of Rajasthan and related it for cases of infantile methemoglobinemia, which may ultimately lead to death. Higher concentration of nitrate causes various other diseases, *e.g.* stomata's and recurrent respiratory tract infections etc. Based on potential nitrate toxicity studies in view of above health effects, the whole region was classified into five water quality zones. Zone–I water contained nitrate less than or equal to 20 mg/l. It was considered safe zone. No ill health effect was observed. Zone–II drinking water had nitrate concentrations ranging between 20 to 50 mg/l. Here cyanosis diarrhea and stomata's were observed among the population due to use of this water. It was termed as mild problematic zone. Zone–III drinking water contains nitrate concentration ranging between 50 to 100 mg/l (maximum ICMR permissible limit is 100 mg/l). Many cases of methemoglobinemia and other disease were reported for this region. This zone was termed as moderately problematic. Drinking water of zone-IV contained nitrate ranging between 100-200 mg/l. People of this area were suffering with diarrhea and methemoglobinemia and increase in nitrate concentration could further aggravate the diseases and this was highly problematic zone.

Khare Rajesh monitored the physico-chemical parameters of drinking water of Bhopal city. He reported, after investigating samples from various sources that TDS varied from 262 to 972ppm. Total hardness from 280 to 980ppm, alkalinity from 302 to 596ppm and chloride content varying from 21 to 471ppm, phosphate was negligible. 50% of samples had nitrate above 42ppm, the range being 9.69 to 60.32ppm, magnesium was between 13 and 446ppm.

1.5.2 Ground water for Irrigation

Ramkumar and coworkers determined the groundwater quality in parts of Vedaraniyam region, Tamil Nadu, India. Irrigation water quality of the groundwater samples fall in the field of C4S4 and C3S1 indicating high to median salinity and high to low sodium water type which can be moderately suitable for irrigation purposes.

Contamination of groundwater in Tirupur region, Tamil Nadu, India has been reported by Karuppapillai and cowerkers in various places. Irrigation parameters were also evaluated by them sodium percentage about 48% of groundwater samples were within the permissible limits, 26% were good, 24% were doubtful and 2% were excellent. A sodium percentage more than 60% is considered unsafe for irrigation. Residual Sodium Carbonate (RSC), 79%, 11% and 10% of the groundwater samples fall in the categories of good, doubtful and unsuitable respectively. Most of the sample falls in the field of C3S1, indicating high salinity and low sodium water, which can be used for almost all types of soil with little danger of exchangeable sodium.

The seasonal variations of the chemical budget of ions in the proximity of River Cooum, Chennai, India were determined from the hydrochemical investigation of the groundwater by Giridharan and cowerkers. The quality of the groundwater in the study area has been assessed using Percent sodium, SAR and Wilcox diagrams.

The lower Varuna River basin in Varanasi district in the central Ganga plain is studied by Raju and cowerkers. The residual sodium carbonate, have revealed that all ground waters were in general safe for irrigation. Permeability index indicates that the groundwater samples are suitable for irrigation purpose. The electrical conductivity values ranges from 500 to 2350 μ S/cm with an average of 948 μ S/cm. The calculated values of percent sodium range 10.13 to 53.83 with an average of 30.7. Wilcox diagram revealed that, out of 75 samples in the study area 28 per cent belongs to excellent to good category, followed by 69 per cent samples with good to permissible category and 3 per cent to doubtful to unsuitable categories for irrigation. The SAR values in the study area ranged from 0.36 to 4.31 with an average of 1.6. One sample fell in the C4S1 (very high salinity with low alkalinity hazard) category. One sample falls in C3S2 type (high salinity and medium sodium hazards). The calculated RSC values of groundwater samples of the study area ranged from 0.2 to 3.3 with an average of 0.92. The groundwater resources contaminated with high levels of nitrate (>45 mg/l) are an environmental hazard.

Groundwater qualities of coastal aquifers in the Chennam-Pallippuram Panchayath of Alappuzha district, Kerala have been extensively monitored by Manjusree and cowerkers. It has been found that about 83% of the samples fall within C1S1 category, about 83% of the samples of January, 91% of March and 83% of May belong to the excellent category.

1.6 Significance of statistical correlation for physico-chemical parameters

The groundwater quality of shallow coastal aquifers of Rameswaram Island in India is studied by Ramachandramoorthy and cowerkers. The Karl Pearson correlation matrix has approved the maximum influence of calcium and chloride on the total dissolved solids.

Mianping Zheng and coworkers have carried out scientific investigation of salt lake on the Qinghai-Tibet Plateau. They observed negative correlation between the pH and salinities of the lakes.

Pradhan and coworkers studied the influence of this modernization activity on Devi estuary and to understand the quality of the Devi estuarine water in Orissa. The application of the multivariate statistics and principal component analysis to the data sets have indicated three factors each during the summer and winter seasons influencing the water to the extent of 77 and 80% respectively.

The physico-chemical characterization of Kucuk Menderes River coastal wetland, Selcuk–Izmir, Turkey was done by Somay and cowerkers and found a good positive correlation between Na and Cl ions ($r^2 = 0.9$) and between SO_4^{2-} and Cl^- ions ($r^2 = 0.9$) and poor correlation between Ca and HCO_3 ions ($r^2 = 0.02$), Ca and Cl ions ($r^2 = 0.3$).

The study was undertaken by Shyamala and cowerkers to characterize the physicochemical nature of groundwater in Telungupalayam village in Coimbatore city by taking water samples from five different stations. A systematic calculation was made by them to determine the correlation coefficient 'r' amongst the parameters and the significant values of the observed correlation coefficient between the parameters.

Ammar Tiri and cowerkers calculated hydrochemical analysis and assessment of water quality in koudiat medouar reservoir, Algeria. The analytical results showed that the EC has good positive correlation with Ca, Mg, Na, K, Cl, SO_4 and HCO_3.

Hajalilou and cowerkers investigated hydro geochemical factors and groundwater quality assessment in Marand Municipality, northwest of Iran. Correlation matrix results indicated that the main type of groundwater was bicarbonate with influential role of alkaline earths. The correlation matrix between chemical variables indicated that there was the highest correlation between HCO_3^- and Mg^{2+} ions (more than 80%) in the spring samples. The other most important correlation has been revealed between Ca^{2+} and Cl^- as 70%.

The application of multivariate statistical analysis was carried out by Voudouris and cowerkers to analyze hydrochemical data in the assessment of groundwater hydrochemistry, especially in situations where numerous samples were available. Simple and multiple regression, factor, and trend-surface analyses were applied by them in order to examine the importance of each parameter, investigate correlations among them, and separate them into groups.

Hydrochemical investigation of water from the Pleistocene wells and springs, Jericho area, Palestine was done by Khayat and coworkors. Mg, K and Na showed good linear correlations with the calculated TDS values while Ca, SO_4 and HCO_3 showed relatively low correlations and nitrate showed very poor correlation.

Silva-Filho and cowerkers studied a geochemical characterization of the coastal aquifer from the oceanic region of Niterói, Rio de Janeiro, South east Brazil by using correlation matrices and factor analysis. The main contributors to the groundwater salinity were Na^+, Ca^{2+}, Mg^{2+} and Cl^-, that shows significant correlation with EC.

Geochemical data on dissolved major constituents in groundwater samples from Manukan Island, Sabah, Malaysia was studied by Aris AZ and cowerkers. There was a positive correlation of Ca^{2+} with Mg^{2+} in the groundwater of the area. These elements, Ca^{2+} and Mg^{2+}, correlated positively with alkalinity, TDS, EC, K^+ and HCO_3^-. The positive correlation between Ca^{2+} and Mg^{2+} (0.536) were attributed to the precipitation of aragonite, dolomite and calcite. Potassium showed positive and significance correlation with all the analyzed parameters except HCO_3^-, with a significance level of more than 0.01. There was positive and significant correlation of K^+ with both Cl^- and SO_4^{2-}. The correlation between sodium and all parameters except HCO_3^- (-0.586) shows a positive correlation.

Groundwater qualities of coastal aquifers in the Chennam-Pallippuram Panchayath of Alappuzha district, Kerala have been extensively monitored by Manjusree and cowerkers. The highly competitive relationship: SO_4^{2-} with Cl^- (0.58) has significant correlation; Ca^{2+} with Na^+ (0.23); Mg^{2+} with Na^+ (0.15); SO_4^{2-} with HCO_3^- (0.32); and Ca^{2+} with K^+ (0.35) have low positive correlation. The affinity ions relationship: Na^+ with Cl^- (0.41); Na^+ with HCO_3^-(0.46); K^+ with HCO_3^- (0.51) have significant positive correlation. Ca^{2+} with SO_4^{2-} (0.29); Mg^{2+} with SO_4^{2-}(0.05) have low positive correlation.

1.7 Objectives of the present study

In view of the little information available in the literature about the hydrogeochemical characterization and statistical correlation studies on physico-chemical parameters of springs/hand pumps of Almora district. It was considered worthwhile to conduct the present study with the following objectives:

i. To study the major inorganic cationic and anionic behavior of the spring/hand pump water.
   - Major cations: Ca^{2+}, Mg^{2+}, Na^+, K^+
   - Major anions: HCO_3^-, Cl^-, SO_4^{2-}, NO_3^-

ii. To make hydrogeochemical characterization of hill springs/hand pumps.

iii. To study the quality of spring/hand pump water in terms of potability and irrigation.
   - CWQI (Chemical water quality indices)
   - HPI (Heavy metal pollution indices)
   - Irrigation parameter indices

iv. To establish the statistical correlation matrix between major anions and cations by using Karl's Pearson correlation coefficient.

v. To assess the utility of correlation matrix in terms of prediction of respective parameters with the help of regression equations.

References

Green Wood NN and Earnshaw A, Chemistry of the elements. Pregamon Press Oxford, 726, 1984.

Mellor JW, A comprehensive treatise on inorganic and theoretical chemistry: Hydrogen and the composition of water. Longmans Green, London 1(3), 122-146, 1922.

Franks F, Introduction–water, the unique chemical 1(1), water, a comprehensive treatise in 7 volumes, Plenum Press, New York, 1972-82.

Das RC, water and unique and important substances. In proceeding of the symposium of Environment aspects of water resources of Orissa, 30 July 1988 OS PC PB, Bhubaneswar.

Mishra PC, Fundamentals of air and water pollution , Ashish Publishing House 8/81, Punjabi Bagh, New Delhi 110026, 75, 65, 1990.

Baumgartner A and Reichel E, Carbon in the fresh water cycle, in the global carbon cycle (Bolim B, Degens T, Duvigenaud P and Kemp, D.ed) Wiley, 319, 1979.

New Webster's Dictionary of the English language (deluxe encyclopedic edition) manufactured in the United States of America 5(a) 219, 5(b) 737, 1981.

Eldon DE, Bradley FS, Environmental Science. A study of interrelationships (fourth edition) Wm.C. Brown publisher p501.

Singh A, Physico-chemical investigation and study of environmental pollution states of water sources at 'Dahod' Distt. (Raisen) with special reference to its prospective features Thesis submitted to Barkatullah University, Bhopal, 1998.

Goel PK, Water Pollution, cause, effects and control. New Age International Ltd. Publishers, 2-5, 1997.

Eldon DE, and Erner RA, The global water cycle. Prentice-Hall, Englewood cliffs, N.J., p.397, 1987.

Meyback M, concentration des eaux fluviales in elements kajeurs et apports en solution aux oceans *Rev. Geol. Dyn. Phys.* 21(3), 215-246, 1979.

Living Stone DA, chemical composition of river and lakes, in M. Fleischer (Ed.), Data of geochemistry 6th ed. U.S. *Geol. Surv. Prof. Paper* 440, 1963.

Freeze RA and Cherry JA, Groundwater. New Jersey, Prentice Hall, 1979.

Ophori DU and Toth J, Patterns of groundwater chemistry, Ross Creek Basin, Alberta, Canada. *Groundwater* 27(1), 20– 26, 1989.

Domenico PA and Schwartz FW, Physical and chemical hydrogeology. Wiley, New York, 410-420, 1990.

Cederstorm DJ, Genesis of groundwater in the coastal plain of Virginia. *Environ Geol*, 41, 218–245, 1946.

Ahmed AA and Ali MH, Hydrochemical evolution and variation of groundwater and its environmental impact at Sohag, Egypt. *Arab J Geosci, 4,* 339–352, DOI 10.1007/s12517-009-0055-z, 2011.

Elewa SA, Effect of the construction of Aswan High Dam on the groundwater in the area between Qena and Sohag, Nile Valley, Egypt. Ph.D. Thesis, Fac. Sci., Assiut Univ., Egypt 2004.

Abdel Rahman AA, Hydrogeological and geophysical assessment of the reclaimed areas in Sohag, Nile Valley, Egypt. Ph.D. Thesis, Geol. Dept., Fac. Sci., Ain Shams University, Cairo, Egypt, 2006.

Prasanna MV, Chidambaram S, Kumar GS, Ramanathan AL and Nainwal HC, Hydrogeochemical assessment of groundwater in Neyveli Basin, Cuddalore District, South India. *Arab J Geosci,* 4, 319–330, DOI 10.1007/s12517-010-0191-5, 2011.

Carson Rachel, Silent Spring, Boston, Houghton Mifflin, 1962. Daley, Herman E.ed. Economics, ecology and Ethics. San Francisco W.H. Freeman, 1980.

McLaughlin, Dorothy. "Fooling with Nature: Silent Spring Revisited". Frontline. PBS. Retrieved August 24, 2010.

Guendouz A, Moulla AS, Edmunda WM, Zouari K, Shand P and Mamou A, Hydrogeochemical and isotopic evolution of water in the complex terminal aquifer in the Algerian Sahara. *Hydrogeol* J, 11, 483–495, 2003.

Wen X, Wu Y, Su J, Zhang Y and Liu F, Hydrochemical characteristics and salinity of groundwater in the Ejina basin, northwestern China. *Environ Geol,* 48, 665–675, 2005.

Wen XH, Wu YQ and Wu J, Hydrochemical characteristics of groundwater in the Zhangye basin, northwestern China. *Environ Geol,* 55, 1713–1724, 2008.

Edmunds WM, Ma JZ, Aeschbach-Hertig W, Kipfer R and Darbyshire DPF, Groundwater recharge history and hydrogeochemical evolution in the Minqin basin, North West China. *Appl Geochem,* 21, 2148–2170, 2006.

Taheri Tizro A and Voudouris KS, Groundwater quality in the semi-arid region of the Chahardouly basin, West Iran. *Hydrol Process,* **22**, 3066–3078, 2008.

Sikdar PK, Sarkar SS and Palcoudhury S, geochemical evolution of groundwater in the quaternary aquifer of Calcutta and Howrah, India. *J Asian Earth Sci,* **19**, 579–594, 2001.

Apodaca LE, Jeffrey BB and Michelle CS, Water quality in shallow alluvial aquifers, Upper Colorado River Basin, Colorado. *J Am Water Resour Assoc,* **38,** 133–143, 2002.

Tesoriero AJ, Spruill TB and Eimers L, Geochemistry of shallow groundwater in coastal plain environments in the southeastern United States: implication for aquifer susceptibility. *Appl Geochem,* **19**, 1471–1482, 2004.

Moller P, Rosenthal E, Geyer S, Guttman J, Dulski P and Rybakov M, Hydrochemical processes in the lower Jordan valley and in the Dead Sea area. *Chem Geol,* **239**, 27–49, 2007.

Karanth KB, Groundwater assessment, development and management. *McGraw-Hill Publishers, New Delhi*, 1997.

Bhatt KB and Salakani S, Hydrogeochemistry of the upper Ganges River, India. *J Geol Soc India*, **48**, 171–182, 1996.

Tardy Y, Characterization of the principal weathering types by the geochemistry of waters from European and African crystalline massifs. *Chemical Geology*, **7**, 253–271, DOI: 10.1016/0009-2541(71) 90011-8, 1971.

Jalali M, Groundwater geochemistry in the Alisadr, Hamadan, western Iran. *Environ Monit Assess*, **166**, 359–369, 2010.

Fianko JR, Adomako D, Osae S, Ganyaglo S, Kortatsi BK, Tay CK and Glover ET, The hydrochemistry of groundwater in the Densu River Basin, Ghana. *Environ Monit Assess*, **167**, 663–674, DOI 10.1007/s10661-009-1082-7, 2010.

Herlinger JR and Viero AP, Hydrogeochemistry of the Coxilha das Lombas Aquifer, Brazil. *Environ. Chem. Lett.*, **5**, 91-94, DOI: 10.1007/s10311-006-0083-9, 2007.

Sadashivaiah C, Ramakrishnaiah CR and Ranganna G, Hydrochemical Analysis and Evaluation of Groundwater Quality in Tumkur Taluk, Karnataka State, India. *Int. J. Environ. Res. Public Health*, **5(3)**, 158-164, 2008.

Jeen SW, Kim JM, Ko KS, Yum B and Chang HW, Hydro geochemical characteristics of groundwater in a mid-western coastal aquifer system, Korea. *Geosciences Journal*, **5(4)**, 339-348, 2001.

Alagbe SA, Preliminary evaluation of hydrochemistry of the Kalambaina Formation, Sokoto Basin, Nigeria. *Environ. Geol.*, **51**, 39–45, DOI: 10.1007/s00254-006-0302-5, 2006.

Subramani T, Rajmohan N and Elango L, Groundwater geochemistry and identification of hydro geochemical processes in a hard rock region, Southern India. *Environ. Monit. Assess*, **162**, 123–137, DOI: 10.1007/s10661-009-0781-4, 2010.

Holland M and Witthuser KT, Geochemical characterization of karst groundwater in the cradle of humankind world heritage site, South Africa. *Environ. Geol.*, **57**, 513–524, DOI: 10.1007/s00254-008-1320-2, 2009.

Hajalilou B and Khaleghi F, Investigation of hydro-geochemical factors and groundwater quality assessment in Marand Municipality, northwest of Iran: A multivariate statistical approach. *Journal of Food, Agriculture & Environment*, **7(3-4)**, 930 – 937, 2009.

Al-Agha MR and El-Nakhal HA, Hydrochemical facies of groundwater in the Gaza Strip, Palestine. *Hydrological Sciences–Journal–des Sciences Hydrologiques*, **49(3)**, 359-371, 2004.

Khayat S, Hotzl H, Geyer S and Ali W, Hydrochemical investigation of water from the Pleistocene wells and springs, Jericho area, Palestine. *Hydrogeol J*, **14**, 192–202, DOI: 10.1007/s10040-004-0399-0, 2006.

Ramkumar T, Venkatramanan S, Mary IA, Tamilselvi M and Ramesh G, Hydro geochemical Quality of Groundwater in Vedaraniyam Town, Tamil Nadu, India. *Res. J. Environ. Earth Sci.,* **2(1)**, 44-48, 2010.

El-Fiky AA, Hydro geochemical Characteristics and Evolution of Groundwater at the Ras Sudr-Abu Zenima Area, Southwest Sinai, Egypt. *Earth Sci.,* **21(1)**, 79-109, DOI: 10.4197 / Ear. 21-1.4, 2010.

Silva-Filho EV, Barcellos RGS, Emblanch C, Blavoux B, Sella SM, Daniel M, Simler R, Wasserman JC, Groundwater chemical characterization of a Rio de Janeiro coastal aquifer, SE – Brazil. *J. of South American Earth Sci.,* **27**, 100–108, DOI:10.1016/j.jsames.2008.11.004, 2009.

Petalas C, Lambrakis N and Zaggana E, Hydrochemistry of waters of volcanic rocks: the case of the volcano sedimentary rocks of Thrace, Greece. *Water, Air, and Soil Pollution,* **169**, 375–394, 2006.

Karuppapillai A and Krishnan E, Quality Characterization of Groundwater in Tirupur Region, Tamil Nadu, India. *International Journal of Applied Engineering Research,* **5(1)**, 9–24, 2010.

Laluraj CM, Gopinath G and Dinesh Kumar PK, Groundwater Chemistry of Shallow Aquifers in the Coastal Zones of Cochin, India. *Applied Ecology and Environmental Research,* **3(1)**, 133-139, 2005.

Giridharan L, Venugopal T and Jayaprakash M, Evaluation of the seasonal variation on the geochemical parameters and quality assessment of the groundwater in the proximity of River Cooum, Chennai, India. *Environ Monit Assess,* **143**, 161– 178, DOI: 10.1007/s10661-007-9965-y, 2008.

Raju, NJ, Ram P and Dey S, Groundwater Quality in the Lower Varuna River Basin, Varanasi District, Uttar Pradesh. *Journal Geological Society of India,* **73**, 178-192, DOI: 0016-7622/2009-73-2-178, 2009.

Manjusree TM, Joseph S and Thomas J, Hydrogeochemistry and Groundwater Quality in the Coastal Sandy Clay Aquifers of Alappuzha District, Kerala. *J. Geo. Soc. of India,* **74**, 459-468, DOI: 0016-7622/2009-74-4-459, 2009.

Jalali M, Geochemistry characterization of groundwater in an agricultural area of Razan, Hamadan, Iran. *Environ. Geol.,* **56**, 1479-1488, DOI: 10.1007/s00254-008-1245-9, 2009.

Rao NS, Nitrate pollution and its distribution in the groundwater of Srikakulam district, Andhra Pradesh, India. *Environ Geol,* **49,** 413–429, 2006.

Fianko JR, Nartey Vincent K and Donkor A, The hydrochemistry of groundwater in rural communities within the Tema District, Ghana. *Environ Monit Assess,* DOI: 10.1007/s10661-009-1125-0, 2009.

Ramachandramoorthy T, Sivasankar V and Subramanian V, The seasonal status of chemical parameters in shallow coastal aquifers of Rameswaram Island, India. *Environ. Monit. Assess,* **160**, 127–139, DOI: 10.1007/s10661-008-0663-1, 2010.

Avvannavar SM and Shrihari S, Evaluation of water quality index for drinking purposes for river Netravathi, Mangalore, South India. *Environ Monit Assess,* **143 (1-3)**, 279-290, DOI: 10.1007/s10661-007-9977-7, 2008.

Abdul Jameel A, Evaluation of drinking water quality in Truchirapalli Tamil Nadu. *Ind. Envir.,* **44(2)**, 168-172, 2002.

Kumar, S, Gupta AB and Gupta S, Need for revision of nitrates standards for drinking water: A case study of Rajasthan. *Ind. J. Env. Hlth.,* **44(2)**, 168-172, 2002.

Khare Rajesh, Ph.D. thesis in chemistry, Barkatullah University, Bhopal, 1997.

Zheng M and Liu X, Hydrochemistry of Salt Lakes of the Qinghai-Tibet Plateau, China. *Aqua. Geochem.,* **15**, 293–320, DOI: 10.1007/s10498-008-9055-y, 2009.

Pradhan UK, Shirodkar PV and Sahu BK, Physico-chemical characteristics of the coastal water off Devi estuary, Orissa and evaluation of its seasonal changes using chemometric techniques. *Current Science,* **96(9)**, 1203-1209, 2009.

Somay AM, Gemici U and Filiz S, Hydro-geochemical investigation of Kucuk Menderes River coastal wetland, Selcuk–Izmir, Turkey. *Environ Geol,* **(16)**, DOI: 10.1007/s00254-007-0972-7, 2007.

Shyamala R, Shanthi M and Lalitha R, Physico-chemical Analysis of Borewell Water Samples of Telungupalayam Area in Coimbatore District, Tamilnadu, India. *E-Journal of Chemistry,* **5(4)**, 924-929, 2008.

Tiri A and Boudoukha A, Hydrochemical Analysis and Assessment of Surface Water Quality in Koudiat Medouar Reservoir, Algeria. *European Journal of Scientific Research,* **41(2)**, 273-285, 2010.

Voudouris K, Panagopoulos A and Koumantakis J, Multivariate Statistical Analysis in the Assessment of Hydrochemistry of the Northern Korinthia Prefecture Alluvial Aquifer System (Peloponnese, Greece). *Natural Resources Research,* **9(2)**, 135-140, 2000.

Aris AZ, Abdullah MH and Kim KW, Hydro-geochemistry of groundwater in Manukan Island, Sabah. *The Malaysian Journal of Analytical Sciences,* **11(2)**, 407–413, 2007.

Muhammad Al-Khatib, Husam Al-Najar, Hydro-geochemical Characteristics of Groundwater beneath the Gaza Strip *Journal of Water Resource and Protection,* **3**, 341-348 DOI:10.4236/jwarp.2011.35043, 2011.

2

Materials and Methods

All chemicals used are of AR grade and double distilled water is used to prepare solutions. The temperature, electrical conductivity (EC), TDS, Eh and pH are measured by digital portable kit (MAC model no. MSW-551). Standard methods have been used to estimate chemical parameters *viz.* sodium, potassium, calcium, magnesium, nitrate, chloride, total dissolved salt and total dissolve solid hardness, sulfate, bicarbonate. Methods employed to calculate CWQI (chemical water quality index), HPI (heavy metal pollution index), correlation coefficient 'r', magnesium content, sodium content, sodium percentage (Na%), sodium adsorption ratio (SAR), residual sodium carbonate (RSC) and permeability index (PI) are the part of this chapter. Application of Piper tri-linear diagram, Durov diagram and Gibbs diagram to interpret hydro-geochemistry and Wilcox diagram to interpret salinity for irrigation of springs have also been given in this chapter.The water of springs was collected from four different regions *viz.* Almora city, Someshwar region, Majkhali region, Jalna region, Panuanaula region and collected water samples from Hand Pump also. These samples were collected seasonally pre-monsoon, monsoon and post-monsoon during 2007 and 2008.

The collection, preservation and analysis of water samples were done following standard methods as prescribed by APHA, Manicaskam, Trivedi and Goel and NEERI.

The water samples analyzed for various physico-chemical parameters are temperature, pH, EC, sodium, potassium, calcium, magnesium, nitrate, chloride, total dissolved salt (TDS), free CO_2, hardness, sulfate, bicarbonate, and heavy metals.

Reagents used for the present investigation was of AR grade and double distilled water was used for the preparation of various solutions.

2.1 Methods Used For the Analysis of Various Parameters

Direct measurements were made at each site with a digital portable kit (Model MAC MSW-551) set of probes, giving readings for environment sensitive index parameters such as temperature, electrical conductivity (EC), TDS, and ORP. For determination of DO sample was preserved by adding one ml of each manganous sulfate and alkaline potassium iodide solution using separate pipettes. Free CO_2 was measured within 5 hours. Other parameter was preserved by adding toluene and brought to the laboratory for detailed chemical analysis. The water samples were kept inside the freezer within 24 hours.

2.1.1 Temperature

Impinging solar radiation and the atmospheric temperature bring about interesting spatial and temporal thermal changes in natural waters which manifest in setting up of convection currents and thermal stratification. Discharge of heated effluents also brings about thermal changes in natural waters (thermal pollution). Temperature is basically an important factor for its effects on chemical and biological reactions in water. A rise in temperature of water accelerates chemical reactions, reduces solubility of gases, amplifies taste and odour, and elevates metabolic activity of organisms.

2.1.1.1 Procedure

Temperature was measured by using digital portable kit (Model MAC MSW-551). It was measured by connecting the temperature rod into temperature socket and the master control selector switch was selected to temperature mode. The bottom of the sensor was touched with water of the sample. The instrument was switched on and allowed to warm up for 15 minutes before measuring sample temperature. Temperature was than recorded.

2.1.2 Hydrogen Ion Concentration (pH or Potentia hydrogenii)

pH is a scale of intensity of acidity or alkalinity and measuredthe concentration of hydrogen ions in water. pH value is expressed as the negative logarithm of the hydrogen ion concentration. Thus pH 7 indicates neutral water, pH 7 to 14 alkaline and pH below 7 acidic.

pH of natural waters varies around 7, generally over 7 (*i.e.* alkaline) due to presence of sufficient quantities of carbonates. It increases during day largely due to photosynthetic activity (consumption of carbon dioxide), whereas decreases at night due to respiratory activity. Factors like exposure to air; temperature and disposal of industrial wastes *etc.* also bring about changes in pH.

2.1.2.1 Procedure

pH was measured by using digital portable kit (Model MAC MSW-551). The operation was performed according to instruction manual of the pH meter to be used. Wash the electrode with distilled water and connect it with the pH meter.

Dip the electrode in the buffer of pH 4.0 and pH 9.2 and move the temperature compensation knob to the temperature of the buffer. In doing so, the meter is calibrated. Put selector switch to zero, wash the electrode with distilled water and dip in the sample. Adjust temperature compensation knob to the temperature of the sample. Read the meter for the pH of the sample.

2.1.3 Electrical Conductivity (EC)

Pure water is a conductor of electricity. Acids, bases and salts in water make it relatively good conductor of electricity. Such substances are called electrolytes. Thus higher concentration of electrolytes in water, the more is its electrical conductance.

Conductance is the reciprocal of the resistance involved and the unit of measure of conductance is reciprocal ohm designated as mho or Siemens.

Conductivity varies with temperature. The conductance of distilled water ranges from 1 to 5 μmho (μS). Waters having up to 20 μmho (μS) conductance are considered to be suitable for irrigation.

2.1.3.1 Procedure

Study carefully the operation manual of the digital portable kit (Model MAC MSW-551). Note the temperature of the sample and adjust the temperature compensation knob of the conductivity meter to the temperature of the sample. Dip the conductivity cell in the sample and note the deflection (dial reading).

2.1.4 Redox Potential or Oxidation-Reduction Potential (ORP)

Redox potential is the measure of oxidation or reducing power of the water. The value of redox potential gives a picture of oxidation and reduction processes going on in water; a low value of redox potential indicates high reduction while a high value of it indicates high oxidation. For better aerobic treatment of waste water redox potential should be high (+200 to +600 mV) while for anaerobic treatment it should be low (-100 to 200 mV).

2.1.4.1 Procedure

Digital portable kit (Model MAC MSW-551) has a facility to measure potential with a pH meter. ORP can be measured by replacing he glass electrode of pH meter with platinum electrode. Set the selector switch to mV. Immerse the electrode in the sample and read the potential on mill volt scale.

2.1.5 Total Dissolved Solids (TDS)

A large number of salts are found dissolve in natural water, the common ones are carbonates, bicarbonates, chlorides, sulfates, phosphates and nitrates of calcium, magnesium, sodium, potassium, iron and manganese etc. A high content of dissolve solids elevates the density of water, influences osmoregulation of freshwater organism and reduces solubility of gases (like oxygen) and utility of water for drinking, irrigational, and industrial purposes.

2.1.5.1 Procedure

Turn the knob of digital portable kit (Model MAC MSW-551) to TDS. Dip the electrode in sample and read the deflection.

2.1.6 Dissolved Oxygen (DO)

DO is a very important parameter of water quality and is an index of physical and biological processes going on in water. There are two main sources of dissolved oxygen in water *(i)* diffusion from air *(ii)* photosynthetic activity within water. Diffusion of oxygen from air to water is a physical phenomenon and depends upon solubility of oxygen which is influenced by factors like temperature, water movements and salinity *etc.* Photosynthetic activity out by autotrophs (mainly phytoplankton in water) and depends upon autotroph population, light conditions and available gases etc.

Non-polluted surface waters are normally saturated with dissolve oxygen while presence of oxygen demanding pollutants (like organic wastes) causes rapid depletion of dissolve oxygen from water. Oxydizable inorganic substances like hydrogen sulphide, ammonia, nitrates, ferrous iron *etc.* also causes decreases in dissolved oxygen.

Winkler's method

2.1.6.1 Reagents

a. **Sodium thiosulfate solution (0.025 N):** Dissolve 6.205 g of sodium thiosulfate in previously boiled distilled water and make up the volume to 1 l. Add a pallet of sodium hydroxide as a preservative. Keep the solution in colored bottle.

b. **Manganous sulfate solution:** Dissolve 100 g of manganous sulfate in 200 ml of previously boiled distilled water and filter the solution.

c. **Alkaline potassium iodide solution:** Dissolve 100 g of potassium hydroxide and 50 g of potassium iodide in 200 ml previously boiled distilled water.

d. **Starch indicator:** Dissolve 1 g of starch in 100 ml of warm distilled water and add a few drops of toluene or formaldehyde as preservative.

e. **Sulfuric acid:** Concentrated sp.gr.1.84; 18 M.

2.1.6.2 Procedure

Take a stoppered BOD bottle of known volume (100 – 300 ml) and fill it with sample avoiding and bubbling. No air should be trapped in bottle after the stopper is placed. Open the bottle and pour in it 1 ml of each manganous sulfate (reagent B) and alkaline potassium iodide (reagent C) solutions using separate pipettes. If the volume of sample is over 200 ml, add 2 ml of each reagent instead of 1 ml. A precipitate will appear. Place the stopper and shake the bottle thoroughly. Sample at this stage can be stored for few days. Add 2 ml of sulfuric acid (reagent E) and shake thoroughly to dissolve the precipitate. Transfer gently whole content in a conical flask. Put a few drops of starch indicator (reagent D). Titrate against sodium thiosulfate solution (reagent A) and note the end point when initial blue color turns to colorless.

2.1.6.3 Calculation

If whole content is used for titration:

$$DO(m/l) = \frac{V_1 \times 8 \times 1000}{V_2 - V_3}$$

If a fraction of the contents is used for titration:

$$DO(m/l) = \frac{P \times 100}{S}$$

Where, DO = dissolved oxygen; V_1 = volume of titrant (ml); N = normality of titrant (0.025); V_2 = volume of sampling bottle after placing the stopper (ml); V_3= volume of manganous sulfate + potassium iodide solutions added (ml); and V_4= volume of fraction of the contents used for the titration (ml).

To obtain the value of DO in ml/l divide the DO in mg/l by 1.43.

2.1.7 Free carbon dioxide (Free CO_2)

Rain water according to its solution equilibrium with the atmosphere air and the absorption coefficient of water for carbon dioxide, contains about 0.6 mg CO_2 per litre. When precipitate water percolates through the soil, additional CO_2 is dissolve out of soil air. Ground waters are extra rich in CO_2.

Free CO_2 dissolve in water is the only source of carbon that can be used in photosynthetic activity of aquatic autotrophs. In the absence of free CO_2, the bicarbonates are converted into carbonates releasing CO_2 which is utilized by autotrophs, thus making the water more alkaline.

2.1.7.1 Reagents

a. **Sodium hydroxide solution (0.2272 N)**: Dissolve 0.909 g of sodium hydroxide in CO_2 free (boiled and cooled) distilled water and make the volume 1 litre. standardize the solution.

b. **Phenolphthalein indicator**: Dissolve 1 g of phenolphthalein in 100 ml of ethyl alcohol and add 100 ml of distilled water. Add NaOH solution (reagent A) drop by drop until a faint pink colour appears.

2.1.7.2 Procedure

After collection analyze the sample as soon as possible. Take 50 ml of sample in a flask and add 2-3 drops of phenolphthalein indicator (reagent B). If the colour turns pink, free CO_2 is absent in a sample. If the sample remains colorless, titrate it against sodium hydroxide solution (reagent A) until pink color appears (end point).

2.1.7.3 Calculation

$$\text{Free } CO_2 \text{ (mg/l) or ppm} = \frac{V_t \times 1000}{V_8}$$

Where, V_t = volume of titrant (ml); and V_s = volume of titrant of sample (ml).

2.1.8 Alkalinity

A number of bases *viz.* carbonates, bicarbonates, hydroxide, phosphates, nitrates, silicates and borates contribute to the alkalinity in natural waters however in natural waters carbonates, bicarbonates and hydroxides are considered to be the predominant bases. Thus alkalinity may be expressed as total alkalinity of alkalinity due to individual bases. In natural waters most of the alkalinity of caused due to CO_2. Natural waters with high alkalinity are generally rich in phytoplankton, especially blue greens. In highly productive waters the alkalinity ought to be over 100 mg/l.

2.1.8.1 Reagents

a. **Sulfuric acid (0.02 N):** Dilute 2.8 ml of concentrated sulfuric acid to 1 l using distilled water. Dilute 200ml of this stock solution (0.1 N) to 1 l using distilled water to prepare 0.02N sulfuric acid titrant. Standardized the solution.

b. **Phenolphthalein indicator**

c. **Methyl orange indicator:** Dissolve 0.1 g of methyl orange in 200 ml of distilled water.

2.1.8.2 Procedure

After collection analysis of sample was done within 5 hour. Take 50 ml of sample in a flask and add 2-3 drops of phenolphthalein indicator. If a slight pink colour appears, phenolphthalein alkalinity is present. Titrate the solution against sulfuric acid (reagent A) until solution becomes colourless (end point). Add 2-3 drops of methyl orange indicator (reagent C) in the same flask and continue to titrate against sulfuric acid (reagent A0 until yellow colour of solution turns orange (end point). Note the reading as **t** which is the volume of titrant used for both the titrations.

2.1.8.3 Calculation

$$\text{Phenolphthalein alkalinity (as } CaCO_{3,} \text{ mg/l)} = \frac{P \times 100}{S}$$

$$\text{Total alkalinity (as } CaCO_{3,} \text{ mg/l)} = \frac{T \times 100}{S}$$

Where, P = volume of titrant used against phenolphthalein indicator (ml); S= volume of sample (ml); and t = total volume of titrant used for the two titrations (ml).

2.1.9 Total carbon dioxide

Total carbon dioxide is the sum of all the three species of CO_2 present in water. It may be used as an index of tropic status.

2.1.9.1 Procedure

Estimate the free CO_2

2.1.9.2 Calculation

Total CO_2 (mg/l) = Free CO_2 (mg/l) + 0.88 (B+C)

Where B = bicarbonate alkalinity (in $CaCO_3$, mg/l); and C = carbonate alkalinity (in $CaCO_3$, mg/l)/2.

2.1.10 Chloride

In natural fresh waters high concentration of chlorides is considered to be an indicator of pollution due to organic wastes of animal origin (animal excreta has high quality of chlorides along with nitrogenous wastes).

2.1.10.1 Principle

Silver nitrate reacts with chloride ions to form silver chloride. The completion of reaction is indicated by the red colour produced by the reaction of silver nitrate with potassium chromate solution which is added as an indicator.

$$AgNO_3 + Cl^- \rightarrow AgCl + NO_3^-$$

$$2AgNO_3 + K_2CrO_4 \rightarrow Ag_2CrO_4 + 2KNO_3$$

2.1.10.2 Reagents

a. **Silver nitrate solution (0.02 N):** Dissolve 3.397 g of silver nitrate in distilled water and dilute to 1 l. store the solution in a dark glass bottle.

b. **Potassium chromate indicator:** Dissolve 10 g of potassium chromate in about 20 ml of distilled water. Add a few drops of 0.02N silver nitrate solution (reagent A) to produce a red precipitate.

2.1.10.3 Procedure

Take 10 ml of sample in a flask and add 5-6 drops of potassium chromate indicator (reagent B). The color of sample becomes yellow. Titrate against silver nitrate solution (reagent A) until a persistent brick red color appears (end point).

$$\text{Chloride (mg/l)} = \frac{V \times N \times 35.457 \times 1000}{S}$$

2.1.10.4 Calculation

Where, V = volume of titrate (ml); N = normality of titrant (0.02); and S = volume of sample (ml).

The salinity of water, on the basis of its empirical relationship with chloride content, may be calculated as follows:

Salinity (mg/l) = 0.03 + 1.805 (chloride in mg/l)

2.1.11 Sulfate

Sulfates are found in appreciable quantity in all natural waters, particularly high in arid and semi-arid regions where natural waters in general have high salt content. Domestic sewage and industrial effluents, besides biological oxidation of reduced sulfur content is high because of industrial and automobile emission, the rain water has high content.

2.1.11.1 Principle

Sulfate ion is precipitated as barium sulfate by adding barium chloride in hydrochloric acid medium. The concentration of the sulfate can be determined from the absorbance of the light by barium sulfate and then comparing it with a standard curve at 420 nm in UV-Visible spectrophotometer (specord 40).

2.1.12 Total hardness

The total hardness of water is the sum of concentration of alkaline earth metal cations present in it. Calcium and magnesium are the principal cations imparting hardness, however, to a lesser extent cations like iron, manganese are also responsible for it. Hardness when caused because of bicarbonates and carbonates of these cations is called temporary hardness since it can be removed by boiling the water. Sulfates and chloride of these cations cause permanent hardness which is not removed by simple boiling of water. In general practice the hardness is measured as concentration of only calcium and magnesium which are far high in concentration over cations.

2.1.12.1 Principle

Calcium and magnesium ions react with EDTA to form soluble complexes and the completion of reaction is indicated by the colour change of a suitable indicator such as Eriochrome Black-T.

$Ca^{2+} + H_2Y^{2-}$ (EDTA) $\rightarrow CaY^{2-} + 2H^-$

$Mg^{2+} + H_2Y^{2-}$ (EDTA) $\rightarrow MgY^{2-} + 2H^-$

MgD(wine red) + H_2Y^{2-} (EDTA) $\rightarrow MgY^{2-} + HD^{2-} + 2H^-$ (blue)

2.1.12.2 Reagents

a. **Ammonium buffer solution:** Dissolve 13.5 g of ammonium chloride in 114 ml of concentrated ammonium hydroxide and add distilled water to make the volume 200 ml.

b. **Erichrome black-T indicator:** Dissolve 0.5 g of erichrome black-T dye in 100 ml of 80% ethyl alcohol.

c. **EDTA solution (0.01 M):** Dissolve 3.723 g of disodium salt EDTA in distilled water to prepare 1 l of solution.

2.1.12.3 Procedure

Take 50 ml of sample in an Erlenmeyer flask and add 1 ml of ammonia buffer solution (reagent A) and 4-5 drop of erichrome black-T indicator (reagent B). Titrate against EDTA solution (reagent C) until the wine color of solution turns blue (end point).

Total hardness (as $CaCO_{3,}$ mg/l=) $\frac{T\times100}{V}$

Where, T = volume of titrant (ml); and V = volume of sample (ml).

2.1.13 Calcium

Calcium is found in great abundance in all natural waters and its source lies in the rocks from which it is leached. Its concentration varies greatly in natural waters depending upon the nature of the basin.

2.1.13.1 Reagents

a. **Sodium hydroxide solution (8%):** Dissolve 8 g of sodium hydroxide in distilled water to 100 ml of solution.

b. **Murexide indicator:** Mix 0.2 g of ammonium purpurate and 100 g of sodium chloride and grind thoroughly.

c. **EDTA solution (0.01):** Dissolve 3.723 g of disodium salt EDTA in distilled water to prepare 1 l of solution.

2.1.13.2 Procedure

Take 50 ml of sample in an Erlenmeyer flask and add 1 ml of sodium hydroxide solution (reagent A) and a pinch of murexide indicator (reagent B). Titrate against EDTA solution (reagent C) until the pink color turns purple (end point).

Calcium (mg/l)= $\frac{T\times400.5\times1.05}{V}$

Where, T = volume of titrant (ml); and V = volume of sample (ml)

To determine the calcium hardness to be expressed in mg/l as $CaCO_3$, employ following formula:

Calcium hardness (mg/l as $CaCo_3$)= $\frac{T\times400.5\times1.05}{V}$

Where, T = volume of titrant (ml); and V = volume of sample (ml)

2.1.14 Magnesium

Like calcium, magnesium is also found in all natural waters and its source too lies in rocks. It is generally in low concentration than calcium.

Magnesium is necessary constituent of chlorophyll without which no ecosystem could operate. Its high content reduces the utility of water for domestic use, while a concentration above 500 mg/l imparts water an unpleasant taste and renders it unfit for drinking purpose. High concentration of magnesium also proves to be diuretic and laxative.

2.1.14.1 Procedure and Calculation

Total hardness and calcium hardness of water as mg/l $CaCO_3$ are determined. From these values magnesium content is calculate as given below:

Magnesium (mg/l)=(T-C) 0.244

Where, T = total hardness (mg/l as $CaCO_3$); and C = calcium hardness (mg/l as $CaCO_3$).

2.1.15 Sodium

This cation occurs generally in lower concentration than calcium and magnesium in fresh waters, and makes its way in water through weathering of rocks. In saline and brackish water its concentration is remarkably high and limits the biological diversity due to osmotic stress. Its salts are highly soluble in water and impart softness (in contrast to hardness).

High sodium content, in the form of chloride and sulfate, makes the water salty in taste and unfit for human consumption. High sodium content in irrigation water causes puddling of soil. As a result water intake of soil is reduced and it becomes hard in which germination of see becomes difficult.

2.1.15.1 Reagents

a. **Standard Sodium solution:** Dissolve 2.5419 g of dried sodium chloride (AR) in distilled water to make 1 l of solution. This stock solution contains 1 g Na/l. Prepare various standard sodium solutions of different strengths (preferably in the ranges of 9 to 1, 0 to 10 and 0 to 100 mg Na/l) by diluting this stock solution with distilled water.

2.1.15.2 Procedure

Read the operation manual of photo meter at hand carefully. Set the filter for reading at 589 nm. Start the compressor and light the burner of flame photometer. Keep the air pressure at 5 lbs and adjust the gas feeder so as to have a blue sharp flame. Feed the standard sodium solution of the highest value in the range and adjust the flame photometer to read full value of emission on the scale. Adjust the zero value of the meter by feeding distilled water. Now feed different standard sodium solutions within the range (*i.e.* 0-1, 0-10 or 0-100 mg Na/l) one by one and record the emission value for each. Plot a standard curve between concentration and emission of standard sodium solutions.

Filter the sample through filter paper and feed it in flame photometer. Note the reading for sample in mg/l by comparing the value with standard curve.

2.1.16 Potassium

This cation occurs in natural waters in far lesser concentration than calcium, magnesium and sodium. It behaves in the water as sodium does. Though found in small amounts it plays a vital role in the metabolism of freshwater environments in the metabolism of freshwater environments and considered to be an important macronutrient.

2.1.16.1 Reagents

a. **Standard Potassium solutions:** Dissolve 1.9064 g of potassium chloride (AR) in distilled water to make 1 l of solution. This stock solution contains 1 g K/l. Prepare various standard potassium solutions of different strengths by diluting this stock solution with distilled water.

2.1.16.2 Procedure

Set the filter of flame photometer for reading at 769 nm and proceed for determination of potassium in sample following the methods described for the determination of sodium. Use standard potassium solutions for preparation of standard curve. Express the result of potassium content in mg/l.

2.1.17 Nitrate

Nitrate is the highest oxidized form of nitrogen and in water its most important source is biological oxidation of nitrogenous organic matter of both autochthonous and allochthonous origin. Domestic sewage and agriculture runoff are the chief sources of allochthonous nitrogenous organic matter. Metabolic wastes of aquatic community and dead organism add to the autochthonous nitrogenous organic matter. Nitrifying bacteria (aminifying bacteria, Nitrosomonas, Nitrobactor) play significant role in oxidation of such organic matter. Certain nitrogen-fixing bacteria (*viz.* Azobactor) and algae (*viz.*, blue-greens like Anabaena, Nostoc) have capacity to fix molecular nitrogen in nitrates. In ground water nitrates may find way through leaching from soil and at times by concentration.

Phenol disulphonic acid method

2.1.17.1 Reagents

a. **Phenol disulphonic acid:** Dissolve 25 g of white phenol in 150 ml of sulfuric acid (concentrated) and further add 85 ml of sulfuric acid (concentrated). Heat for about 2 hours on a water bath, cool, and keep the solution in a dark bottle.

b. **Potassium hydroxide solution (12 N):** Dissolve 336.5 g of potassium hydroxide in distilled water to make the volume 500 ml.

c. **Standard nitrate solutions:** Dissolve 0.722 g of anhydrous potassium nitrate in distilled water to prepare l of stock solution. This stock solution contains 100 mg NO_3/l (or 443 mg NO_3 ions/l). Prepare standard nitrate solutions of various strengths by diluting stock solution with distilled water.

2.1.17.2 Procedure

Take 25 ml of sample in a porcelain basin and evaporate it to dryness on a hot water bath. Add 0.5 ml of phenol disulphonic acid (reagent A) to the residue and dissolve the latter with the help of a glass spatula. Add 5 ml of distilled water and 1.5 ml of potassium hydroxide solution (reagent B). Stir for through mixing. Take the supernatant of yellow colour and read its absorbance (S) on spectrophotometer at 410 nm. Use process the standard nitrate solutions (reagent C) in similar manner and note the absorbance for each. Plot a standard curve between absorbance and concentrations of various standard solutions. Deduce the value of nitrate and nitrogen in the sample by comparing the absorbance of sample (S) with the standard curve and express the result in mg/l.

2.1.18 Heavy Metals (Mn, Fe, Cu, Zn,)

Heavy metals were estimated by atomic absorption spectrometer (Model Vario 06) at Agro chemistry Cell of Vivekananda Parvatiya Krishi Anusandhan Sansthan, Almora, Uttarakhand.

2.2 Classification of water chemistry data

Graphical and statistical methodologies were used to classify the water samples into homogeneous groups. These methodologies include the diagrams of Piper, Durov, Gibbs and Wilcox. Other classifications based on the SAR-EC, as well as the total hardness, were performed and compared.

2.2.1 Piper diagram

Based on the four main cations (calcium, magnesium, and sodium + potassium) and the four main anions (bicarbonate, sulfate, chloride and nitrate), Piper proposed a tri-linear diagram that permits the classification of waters, according to Langguth into seven types as shown in Fig. 2.1. The AQUA-CHEM 4.0 computer software was used for plotting this diagram.

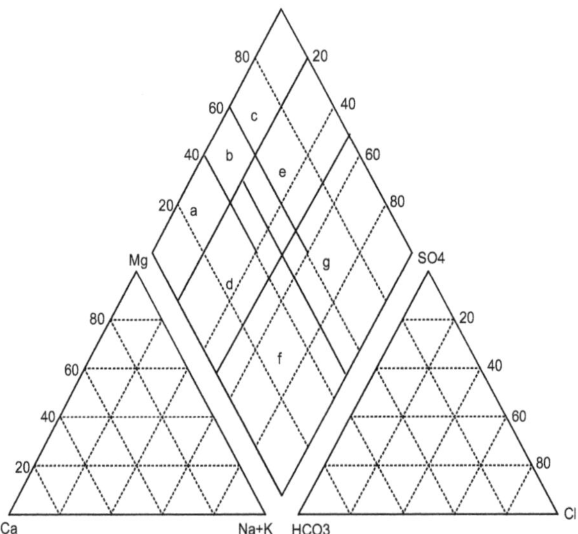

Fig. 2.1: Piper tri-linear diagram

Below a summary about the theory behind the divisions in the diagram is given.

Water Types:

Normal earth alkaline water

a. with prevailing bicarbonate
b. with prevailing bicarbonate and sulfate or chloride
c. with prevailing sulfate or chloride

Earth alkaline water with increased portions of alkalis

d. with prevailing bicarbonate
e. with prevailing sulfate and chloride

Alkaline water

f. with prevailing bicarbonate
g. with prevailing sulfate-chloride

2.2.2 Durov diagram

Durov diagram for the major cations and anions plotted by AQUA-CHEM 4.0 software is illustrated in Fig. 2.2. The fields and lines on the diagram show the classifications of Lloyd and Heathcoat. Below a summary about the theory behind the divisions in the diagram is given:

- **Field (1):** HCO_3 and Ca dominant, frequently indicates recharging waters in limestone, sandstone, and many other aquifers.
- **Field (2):** This water type is dominated by Ca and HCO_3 ions. Association with dolomite is presumed if Mg is significant. However, those samples in which Na is significant, an important ion exchange is presumed.
- **Feild (3):** HCO_3 and Na are dominant, indicates ion exchanged water, although the generation of CO_2 at depth can produced, HCO_3 where Na is dominated under certain circumtance.
- **Field (4):** SO_4 dominates, or anion discriminant and Ca dominant, Ca and SO_4 dominant, frequently indicates recharge water in lava and gypsiferous deposits, otherwise mixed water or water exhibiting simple dissolution may be indicated.
- **Field (5):** No dominant anion or cation indicates water exhibiting simple dissolution or mixing.
- **Field (6):** SO_4 dominant or anion discriminant and Na dominant; is a water type that is not frequently encountered and indicates probable mixing influences.
- **Field (7):** Cl and Na dominant are frequently encountered unless cement pollution is present. Otherwise the water may result from reverse ion exchange of Na-Cl waters.

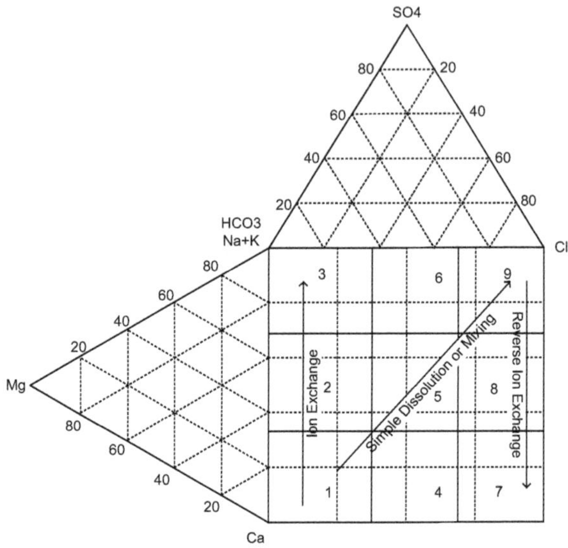

Fig. 2.2: Durov diagram for the major cations and anions.

- **Field (8):** Cl dominant anion and Na dominant cation, indicate that the ground waters be related to reverse ion exchange of Na- Cl waters.
- **Field (9):** Cl and Na dominant frequently indicate end-point waters.

2.2.3 Gibbs diagrams

Gibbs diagrams that represent the ratios of $Na^+ : (Na^+ + Ca^{2+})$ and $Cl^- : (Cl^- + HCO_3^-)$ as a function of TDS are widely employed to understand the functional sources of dissolved chemical constituents, such as precipitation-dominance, rock-dominance and evaporation dominance. The chemical data of groundwater samples are plotted in the Gibbs diagram as depicted below:

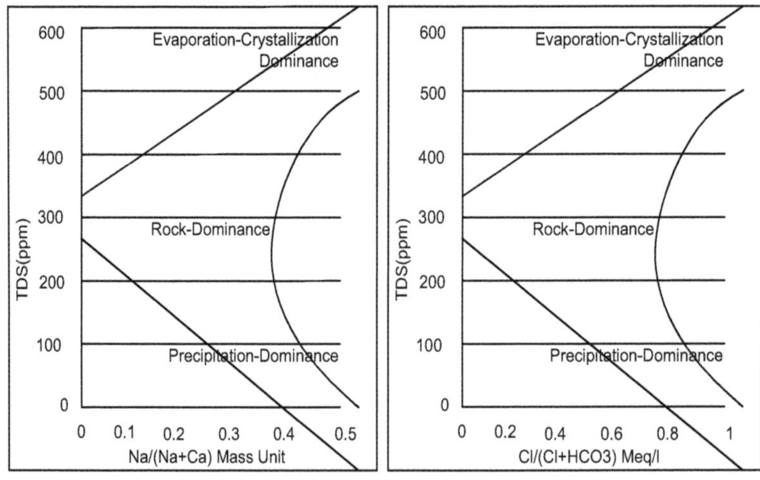

Fig. 2.3: Gibbs diagrams for the cations and anions of groundwater

2.2.4 Statistical analysis

Statistical calculations were conducted using the statistical software packages SPSS and AQUA-CHEM 4.0 computer software. The relation of water quality parameters on each other in samples of water analyzed was determined with regression analysis by determining co-relation co-efficient (r) by using the mathematical formula as given below:

Let x and y be any two variable (water quality parameters in the present case) and n= number of observations.

Then the correlated coefficient (r), between the variable x and y is given by the relation,

$$r = \frac{ns(x.y)-(sx)(sy)}{\sqrt{[f(x).f(y)]}} \quad ...1$$

Where

$$f(x)= ns(x)^2- (sx)^2 \quad ...2$$

$$f(y)= ns(y)^2- (sy)^2 \quad ...3$$

And all the summations are to be taken from 1 to n. If the numerical value of the correlation co-efficient between two variable x and y is fairly large, it implies that these two variables are highly correlated. In such cases, it is feasible to try a linear form.

$$y= Ax+B \quad ...4$$

The constant A and B are determined by fitting the experimental data on the variables x and y to eq.4. According to the well known method of least squares, the value of the constant A and B are given by the relations.

$$A = \frac{ns(x \times y)-(sx)\times(sy)}{ns(x-1/x)^2} \quad ...5$$

$$B=Y-Ax \quad ...6$$

$$\text{where } x=(sx)/n,\ y=(sy/n) \quad ..7$$

All the summations are taken from 1 to n. The co-relation among the different parameters will be true when the values of co-relation co-iefficient (r) are high approaching to one. The statistical interpretation of chemical parameters can provide better tool to reduce the laborious work to study the hydro-geochemistry of springs.

2.2.5 Water quality evaluation

Evaluation of the water quality for domestic uses was based on a comparison of the biological, physical, and chemical parameters in the water of the springs and wells with the drinking water guidelines of WHO. The main parameters of concern were the fecal coliform bacteria, nitrate, and heavy metals. Evaluation of water for agricultural purposes was based on salinity, sodium hazards, soluble sodium percentages, magnesium content, permeability index, and residual sodium carbonate.

To study the trends of water quality in the study area, the Chemical Water Quality Index (CWQI) of each sampling stations and Heavy Metal Pollution Index (HPI) are calculated. Quality indices are useful in obtaining a composite influence of all parameter on overall pollution.

2.2.5.1 Chemical Water Quality Index (CWQI)

The CWQI is based on the sub index functions. To study the trends of water quality in the study area, the CWQI with DO and without DO *viz.* $CWQI_{DO}$ and $CWQI_{WDO}$ of each sampling station is calculated. For calculation of weighed arithmetic $CWQI_{DO}$, the following 12 physico-chemical parameters were taken into account: pH, TDS, total alkalinity, total hardness, dissolved oxygen(DO), chloride (Cl^-), sodium (Na), potassium (K) , calcium (Ca), magnesium (Mg), nitrates (NO_3^-) , Sulphates (SO_4^-) and for $CWQI_{WDO}$ DO has been dropped. It is an established fact that the more harmful a given pollutant is, the smaller is its standard permissible value recommended for drinking water. Therefore, the 'weights' for various water quality characteristics are assumed to be inversely proportional to the recommended standards for the corresponding parameters, that is,

$$W_1 = K/S_i \quad ...1$$

Where Wi is the unit weight and Si is the recommended standard for the ith parameter Pi. The constant of proportionality K in the equation can be determined from the condition:

$$\sum_{i-1}^{n} Wi = K \sum_{i-1}^{n} (1/Si) = 1 \quad ...2$$

The quality rating qi for the ith parameter Pi by the reaction:

$$q_1 = 1000 \backslash (v_1/s_1) \quad ...3$$

Where Vi is the observed value. The sub index (SI)i, the ith parameter Pi is given by

$$(SI)_i = (q_i w_i) \quad ...4$$

For dissolve oxygen, for instant, the ideal value may be taken as 16.4 mg/l (the solubility of oxygen at 0 ^{0}C). The standard for drinking water being 5.0 mg/l, the equation (3) may modify as:

$$qDO = 100[(vDO-11.6)/(5.0-11.6)] \quad ...5$$

Where, vDO is observed value of dissolved oxygen.

Similarly, for pH, the ideal value may be taken as 7.0 and the permissible value 8.5, the equation (3) may modify as:

$$qDO = 100[(vpH-7.0)/(8.5-7.0)] \quad ...6$$

Where, vpH is observed value of pH in the test water.

The overall CWQI can be calculated by aggregating the quality rating (qi) or sub indices, linearly, and taking their weighted mean, i.e.

$$CWQI = \frac{\sum(qiwi)}{\sum wi} \quad ...7$$

The 12 individual sub index functions were combined into one general formula. In table 2.1 is shown a sample calculation of the CWQI for urban spring water of Almora (A1) using equations 1 to 7.

2.2.5.2 Heavy Metal Pollution Index (HPI)

The HPI represents the total quality of spring water with respect to heavy metals. The HPI is based on the weighted arithmetic quality mean method and is developed in to two basic steps. First, by establishing a rating scale for each selected parameter giving weightage to select parameter and, second, by selecting the pollution parameter on which the index is to be based. Iron, Manganese, Copper and Zinc have been monitored for the model index application. The Heavy Metal Pollution Index (HPI) model proposed is given by.

$$HPI = \frac{\sum_{i=1}^{n} WiQi}{\sum_{i=1}^{n} Wi} \quad ...1$$

Table 2.1: Sub Index (SI) function used for different parameters

S. No.	Parameter	Observed value V_i	Standard S_i	Unit w_i	Quality rating q_i	Sub index w_{iq}
1.	pH	-	7.75	0.004	-	-
2.	Total dissolved solid (TDS)	-	500	0.002	-	-
3.	Total Alkalinity	-	120	0.0083	-	-
4.	Total Hardness	-	300	0.0033	-	-
5.	Dissolved oxygen (DO)	-	5	0.20	-	-
6.	Chloride (Cl^-)	-	250	0.004	-	-
7.	Sodium (Na)	-	20	0.05	-	-
8.	Potassium (K)	-	10	0.10	-	-
9.	Calcium (Ca)	-	75	0.013	-	-
10.	Magnesium (Mg)	-	50	0.02	-	-

Contd...

S. No.	Parameter	Observed value V_i	Standard S_i	Unit w_i	Quality rating q_i	Sub index w_{iq}
11.	Nitrate (NO_3	-	45	0.02	-	-
12.	Sulphate (SO_4^{2-})	-	200	0.005	-	-
		-	-	0.43	-	-
Total				$CWQI=\left[\left(\sum wiqi / \sum qi\right)\right]$		

Where Qi is the sub index of the ith parameter. Wi is the unit weightage of ith parameter and n is the number of parameter considered. The Sub-index (Qi) of the parameter is calculated by

$$Q_i = \sum_{i=1}^{n} \frac{\{Mi(-)Ii\}}{(Si\text{-}Ii)} \times 100 \qquad ...2$$

Where Mi is the monitored value of heavy metal of the ith parameter, Ii is the ideal value of ith parameter; Si is the standard value of ith parameter. The sign (-) indicates the numerical difference of the two values, ignoring the algebraic sign. HPI is calculated by using equation (1) and (2) and as given in the table 2.2.

Table 2.2: Heavy Metal Pollution Index (HPI) calculation for spring water

Heavy metals	Mean conc. Value (Mi)	Standard permissible value (Si)	Highest desirable value (Ii)	Unit weightage (Wi)	Sub index (Qi)	Wi*Qi
Fe	...	1,000	100	0.001	...	...
cu	...	1,00	50	0.001	...	...
Mn	...	300	100	0.0033	...	...
Zn	...	15,000	5,000	0.0006	...	...
				HPI=...		

2.2.6 Evaluation of water quality for irrigation uses

The suitability of water for irrigation is determined by its mineral constituents and the type of the plant and soil to be irrigated. Many water constituents are considered as macro or micro nutrients for plants, so direct single evaluation of any constituent of these will not be of great value except if complete analysis of soil and determination of plant need are done. Due to that more generalized criteria, which represent combinations of the different water parameters, were adopted worldwide (i.e. salinity (EC), SAR, SSP, RSC, MH and PI) for the evaluation of water quality for irrigation purposes, and will be used in this work.

2.2.6.1 Salinity

Excess salt increases the osmotic pressure of the soil water and produces conditions that keep the roots from absorbing water. This results in a physiological drought condition. Even though the soil appears to have plenty of moisture, the plants may wilt because the roots do not absorb enough water to replace water lost from transpiration. Based on the EC, irrigation water can be classified into four categories.

2.2.6.2 Sodium hazard

The main problem with high sodium concentration is its effect on soil permeability and water infiltration. Sodium also contributes directly to the total salinity of the water and may be toxic to sensitive crops. The sodium hazard of irrigation water is estimated by the sodium absorption ratio (SAR), which is calculated by the following formula:

$$SAR = Na^{+}/\sqrt{(Ca^{2+} + Mg^{2+}/2)}$$

Where the cations are expressed in meq/L.

Continued use of water having a high SAR leads to a breakdown in the physical structure of the soil. The sodium replaces calcium and magnesium sorbed on clay minerals and causes dispersion of soil particles.

Table 2.3: Classification of irrigation water based on salinity (EC) values.

Lev el	EC (µS/cm)	Hazard and limitations
C1	< 250	Low hazard; no detrimental effects on plants, and no soil buildup expected.
C2	250 - 750	Sensitive plants may show stress; moderate leaching prevents salt accumulation in soil.
C3	750 - 2250	Salinity will adversely affect most plants; requires selection of salt-tolerant plants, careful irrigation, good drainage, and Leaching.
C4	> 2250	Generally unacceptable for irrigation, except for very salt tolerant plants, excellent drainage, frequent leaching, and intensive management.

2.2.6.3 Soluble sodium percentage

Soluble sodium percentage (SSP) is an estimation of the sodium hazard of irrigation water like SAR, but it expresses the percentage of sodium out of the total cations and not as SAR correlating the sodium with the Ca and Mg only. SSP is calculated by the following formula:

$$SSP=\{(Na^{+}+K^{+})/(Ca^{2+}+Mg^{2+}+Na^{+}+K^{+})\}\times 100$$

Where the ionic concentrations are in meq/l.

Table 2.4: Classification of irrigation water based on SAR values

Level	SAR	Hazard
S1	<10	No harmful effects from sodium.
S2	10-18	Appreciable sodium hazard in fine-textured soils of high CEC, but could be used on sandy soils with good permeability.
S3	18-26	Harmful effects could be anticipated in most soils and amendments such as gypsum would be necessary to exchange sodium ions.
S4	>26	Generally unsatisfactory for irrigation.

Wilcox diagram illustrating the chemical analyses of the water samples

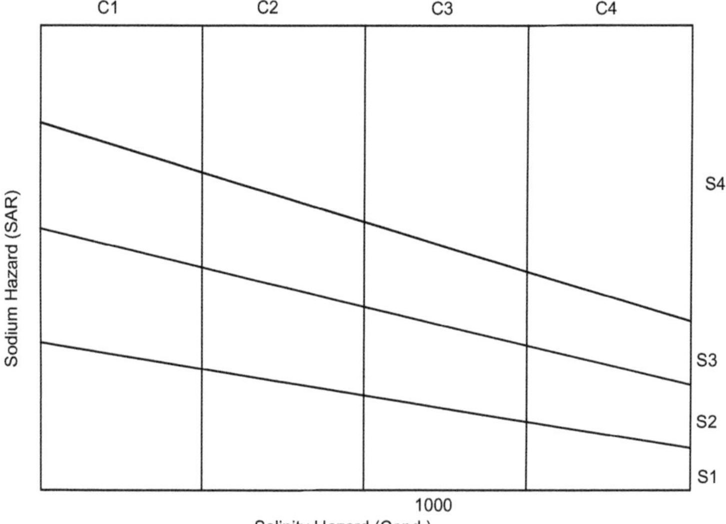

Table 2.5: Classification of irrigation water based on SSP

Water class	SSP	EC (μS/cm)
Excellent	< 20	> 250
Good	20-40	250-750
Permissible	40-60	750-2000
Doubtful	60-80	2000-3000
Unsuitable	> 80	> 3000

2.2.6.4 Residual sodium carbonate

The residual sodium carbonate (RSC) equals the sum of the bicarbonate and carbonate concentrations minus the sum of the calcium and magnesium ion concentrations, where the ions are expressed in meq/L.

$RSC=(HCO_3^{-}+CO_3^{2-}) - (Ca^{2+}+Mg^{2+})$

As RSC increases, much of the calcium and some magnesium are precipitated from the solution when water is applied to soil, increasing the sodium percentages and the rate of sorption of sodium on soil particles which increases the potential for a sodium hazard.

2.2.6.5 Magnesium Hazards (MH)

Mg Hazards calculated by the following formula

$MH=[Mg^{2+}/Ca^{2+}+Mg^{2+})]\times 100$

Magnesium hazard value higher than 50 meq/l is unsuitable for irrigation.

RSC	Hazard
< 0	None.
0-1.25	Low, with some removal of calcium and magnesium from irrigation water.
1.25-2.50	Medium, with appreciable removal of calcium and magnesium from Irrigation water.
>2.50	High, with most calcium and magnesium removed leaving sodium to Accumulate.

2.2.6.6 Permeability Index (PI)

The Permeability index also indicates whether groundwater is suitable for irrigation. Doneen classified irrigation water based on the Permeability index:

$$PI= \frac{Na^{+}+\sqrt{HCO_3^{-}}}{\left(Ca^{2+}+Mg^{2+}+Na^{+}\right)}\times 100$$

Where the concentrations are reported in meq/L. accordingly, water can be classified as Class I, II and III. Class I and II water are categorized as good for irrigation with 75% or more of maximum permeability. Class III water is unsuitable with 25% of maximum permeability.

References

APHA, Standard methods for the examination of water and waste water American Public Health Association, New York, USA, (20th Ed.) 1998.

Manivaskam N, Physico-chemical examination of water, sewage and industrial, Pragati Prakashan Meerut, 1986.

Trivedi PK and Geol PK, Chemical and biological methods for water pollution studies. Env. Publication, Karad, 1986.

Manual on water and wastewater Analysis NEERI publications, 1988.

Instruction Manual for digital portable kit (MSW-551) MAC established; (a) 2, (b, c) 4 (d) 6.

Llloyd JW and Heathcoat JA, Natural inorganic chemistry in relation to ground water. Clarendon Press, Oxford, 1985.

Langguth HR, Die Grundwasserverhältnisse im Bereich des Velberter Sattels, Rheinisches Schiefergebirge. Der Minister für Ernährung, Landwirtschaft und Forsten, NRW, Düsseldorf. (Unpublished) 1966.

Piper AM, Graphical procedure in geochemical interpretation of water analysis. Trans-American Geophysical Union, 25, 914-928, 1944.

Wilcox LV, Classification and use of irrigation waters. US Dept. Agric. Circ. 969. Washington, D.C., USA, 19, 1955.

WHO-World Health Organization, WHO guideline for drinking water quality, 2nd Edition. Vol. 1 – Recommendations. Geneva: 8-29 and 120-130, 1993.

WHO-World Health Organization, Drinking water guidelines, Amman, 1995.

WHO-World Health Organization, Guidelines for drinking water quality, 2nd Edition. Vol. 2 Health criteria and other supporting information. Geneva, 940-949, 1996.

Schoeller H, Les Eaux sutterraines. Masson et cie, 67, Paris, 1962.

Kumar K, Rawat DS and Joshi R, Chemistry of springs in Almora, Central Himalaya, India. *Env. Geol*, **28(2)**, 1-7, 1996.

Sharma C Ramesh, Chauhan, Punam and Bahuguna, Manju, Impact of Tehari dam on aquatic macro invertebrate diversity of bhagirathi, Uttarakhand (India). *Journal of Environ, Science and Engg*, **50(1)**, 41-50, 2008.

Rajmohan N, Elango L, Ramachandran S and natarajan M, Major Ion correlation in ground water of Kancheepuram Region, South India. *Indian J. Environ Hlth*, **45(1)**, 5-10, 2003.

Pathak JK, Alam mohd., and Sharma Shikha, Interpretation of ground water quality using multivariate statistical technique in Moradabad city, western Uttar Pradesh state, India. *E-journal of chemistry*, **5(3)**, 607-619, 2008.

Kowalkowski T, Kbyniewski R, Szpejna J and Buszewski B, Water res., 4974-752, 2006.

Padro REB, Castrillejo Y, Valasco MA and Vaga M, *Anal. lett*, 26, 2617-2639, 1993.

Ubale MB, Farooqi Mazahar, Arif Md. Pathan, Zahher ahmad and Dhule, DG, Regression analysis of ground water quantity data of chiklthana industrial area, Aurangabad (Maharashtra). *Oriental Journal of chemistry*, 17(2), 347-348, 2001.

Sarkar, Mital., Banerjee, Abarana, Pramanick Parth Pratim and Chakraborty, Sumit, Appraisal of elevated fluoride concentration in ground water using statistical correlation and regression study, *J. Indian Chem. Soc.*, 83, 1023-1027, 2006.

Mishra PC, Pradhan KC and Patel RK, Quality of water for drinking and agriculture in and around mines in Keonjhar District, Orrisa. Indian, *J. Environ Hlth.*, 45(3), 273-220, 2003.

Bartarya SK, Hydrochemistry and rock weathering in a subtropical lesser Himalayan river basin in Kumaun, *India. J. Hydrology,* 146, 149-174, DOI: 10.1016/0022-1694(93)90274,1993.

Bhandari NS, Pande H, Kumar S and Pande KK, Nitrate in Ground Waters of Kumaun Hills, Uttarakhand. ENVIS Bulletin: *Himalayan Ecology,* **15(1-2),** 25-28, 2007.

BIS, Bureau of India standard, New Delhi, 1993.

Dinius SH, Design of an Index of Water Quality. *Water resources Bulletin,* **23(5)**, 833-843, 1987.

Kumar K, Rawat DS and Joshi R, Chemistry of spring water in Almora, central Himalaya, India. Environ. Geology, 31(3/4), 150-156, DOI: 10.1007/s002540050174, 1997.

Mariappan P, Vasudevan T and Yegnaraman V, Water quality in drinking water ponds (oorany) in eastern part of sivagongai district, Tamilnadu. J. Ind. water Works Assoc., April-June, 1999.

Majumdar D and Gupta N, Nitrate Pollution of Ground Water and associated human health Disorders. *Indian J. Environ. Hlth* **42(1)**, 28-39, 2000.

Mohan SV, Nithila P, Readdy SJ, Estimation of heavy metal in drinking water and development of heavy metal pollution Index. *J. Environ. Sci. Hlth.* **A 31(2),** 283, 1996.

Murthy R, Venkatamohan GV, Harischandrap S and Karthikeyan J, A preliminary study on water quality of river Tungabhara at kurnool town. *Ind. J. Environ. Prot.,* 14, 8, 1994.

Prasad B and Bose JM, Evaluation of the heavy metal pollution index for surface and spring water near a limestone mining area of the lower Himalayas. *Environ. Geo,* 41, 183-188, DOI: 10.1007/s002540100380, 2001.

Sahu BK, Panda RB, Sinha BK and Nayak A, Water quality index of river Brahnnaniat Rourkela Industrial complex of Orissa. *J. Ecotoxico. Environ. Monit,* 13, 169-175, 1991.

Singh DF, Studies on the water quality Index of some major rivers of Pune, Maharashtra. *Proc. Acad. Environ. Bio,* 1(1), 61-66, 1992.

3

Study Area and Data Compilation

In this chapter, description of study area and all details of data collected are included. Total fifty four springs/hand pumps have been identified for the present study from Almora teshil of Kumaun hills. The regions of Almora Tehsil included are Someshwar, Majkhali, Panuanaula, Jalna and Almora town (urban). The samples were collected seasonally pre-monsoon, monsoon, post-monsoon during two consecutive years 2007 and 2008.

Almora district lies between latitudes 29 37′N and 29.62 N and longitudes 79 40′E and 79.67 E. It has an average elevation of 1,651 meters (5,417 feet) and is located on a ridge at the southern edge of the Kumaun Hills of the Himalaya range.

The geological formation of Almora group consists of fractured mica schists interbedded with thin bends of quartzite, granite, gneiss and nonfoliated granitic rocks, which acts as a barrier to the groundwater flow and results into the origin of springs. The saddle-shaped ridge on which the township is developed forms the recharge area for these springs.

3.1 Description of study area

The study area comprises Someshwar, Majkhali, Panuanaula, Jalna and specially Almora town (urban) regions. There are total fifty four samples including ten hand pumps have been selected for the present investigation (Figure 3.1).

Eight sampling stations are selected in Someshwar region (S), in which all the sampling sites are located within the populated area. Geologically Someshwar lays on Almora Nappe I. The garnetiferous schist and quartzite sequence, interspersed with biotite, muscovite granite, augen and tabular gneiss, is folded into major recumbent fold forming the Almora Nappe I. The northern limit of this Nappe, recognized as north Almora thrust, is a high angle fault heading towards south which becomes vertical near Someshwar. North of Almora, an en echelon fault pattern is seen related to this high angle fault. Agricultural activities are in big bloom,

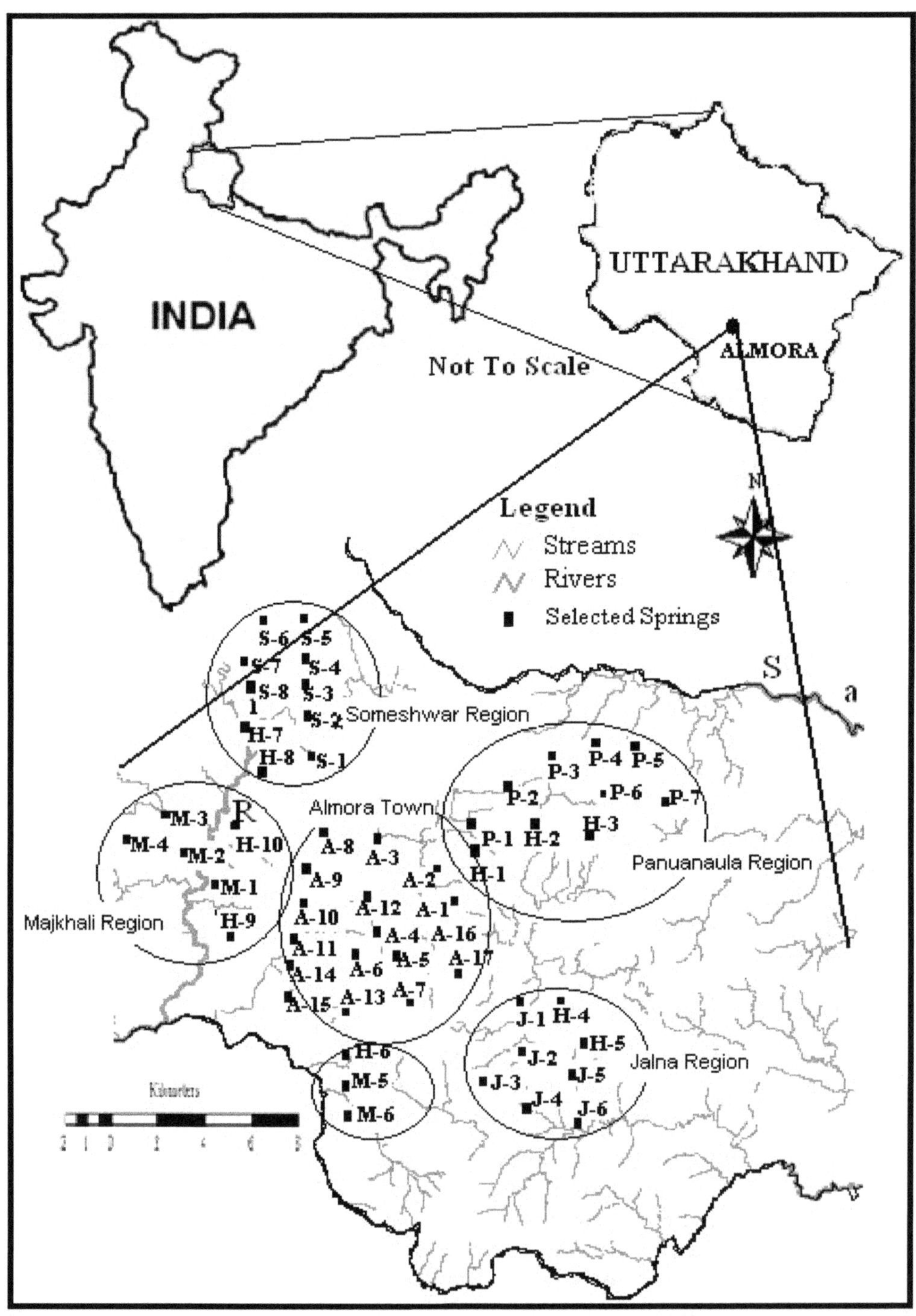

Fig. 3.1: Location map of selected springs.

fertilizers and pesticides are extensively used to meet the agricultural demand in Someshwar region.

Majkhali area lies on the north-dipping southern limb of the synformal Almora Nappe. The mineral assemblage of the formation is as Garnet-mica Schists. In Majkhali region (M) sampling was done from six different springs two of which are located in between the agricultural land and other four springs are from populated area.

Almora town region (A) site reflects the population load of the area. Seventeen sampling springs have been selected for chemical investigation. Agricultural activities are less observed in that area. In Almora town, favorable geological characteristics in the form of fractured mica schists with interbedded thin layers of quartzite are responsible for the occurrence of number of springs within the town boundary. Geologically the town is located on the Almora Nappe, delimited by the North and South Almora Thrust. The Almora Nappe and its rows of detached pieces or klippen are folded with the synclinal cores of pre-Cambrian sedimentary rocks.

Panuanaula is situated in higher hills, covered with less population which is highly involved in agriculture. All sampling sites are located among agricultural lands. Jalna region is located approximately 30 km north to Panuanaula which has the same population distribution and agricultural activities as Panuanaula region. Seven and six samples were taken from Panuanaula (P) and Jalna (J) regions respectively.

Table 3.1: Description of sample identification (ID) and location Name

Sample ID	Location Name	Sample ID	Location Name
	Almora Town Region	S5	Someshwar Market Naula
A1	Sunari Dhara	S6	Kalika Mandir Someshwar Nula
A2	Sunari Naula	S7	Gimkhana Naula Som eshwar
A3	Dauba Naula (A)	S8	Village Palyuda Naula
A4	Dauba Naula (B)		
			Panuanaula Region
A5	Dauba Naula (C)	P1	Panuanaula
A6	Dauba Naula (D)	P2	Panuanaula
A7	Siddh Naula	P3	Moongroo Dhara Panuanaula
A8	Khatyari Naula	P4	Panuanaula
A9	Lower Mall Road Naula	P5	Panuanaula
A10	Campus Field Dhara	P6	Panuanaula
A11	Talla Gururani Naula	P7	Panuanaula
A12	Champa Naula		**Jalna Region**
A13	Kapina Dhara	J1	Jalna
A14	Rani Dhara	J2	Jalna

A15	Chousar Naula	J3	Jalna
A16	Dhara Naula	J4	Jalna
A17	Hand Pump Lower Campus	J5	Jalna
	Majkhali Region	J6	Jalna
M1	Karamal Dhara		**Hand Pumps**
M2	Majkhali Dhara	H1	Hand Pump Maniagar
M3	Majkhali Naula	H2	Hand Pump Barecchina
M4	Majkhali Naula	H3	Hand Pump Petshaal
M5	Khunt Naula	H4	Hand Pump Chitai
M6	Shitlakhet	H5	Hand Pump Mehala
	Someshwar Region	H6	Hand Pump Bhagtola
S1	Maht Gaun	H7	Hand Pump Chinauna
S2	Bharave Mandir Naula	H8	Majkhali Near Birla Mandir
S3	Etaula Dhara	H9	Hand Pump Khoont
S4	Gwalkot	H10	Hand Pump Bimaula

The granite schist in the central part of the crystalline mass is exposed at Panuanaula and Jalna and forms a recumbent fold. This fold may be digitations of the main fold.

These areas also exist within Almora Nappe I. Ten Hand pumps (H) from different regions have been selected for the present study.

Description of the study area and sample identification are tabulated and figured in table 3.1 and figure 3.1 respectively.

3.2 Data presentation

The periodic monitoring of the physico-chemical characteristics of the springs/ hand pumps were done seasonally over the period of two consecutive years, *i.e.* 2007 and 2008. Data tabulation is given in table 3.2 to table 3.7.

Table 3.2: Analytical result of Spring/Hand Pump water samples in the Almora Tehsil area (Pre-monsoon, 2007)

Sample ID Parameter	A1	A2	A3	A4	A5	A6	A7	A8	A9	A10	A11	A12	A13	A14
Temp.	14.5	13.4	15.4	15.4	13.9	12.1	15.8	15.3	13	14.3	14.4	14.5	12.4	16.8
pH	6.7	6.94	6.69	6.68	6.72	6.82	6.39	6.94	6.55	6.6	6.45	6.68	6.48	6.25
Eh	22	4	23	22	22	15	39	5	29	31	44	24	37	52
EC	713	693	600	511	507	456	307	371	404	295	698	681	499	346
TDS	453	440	381	324	323	290	195	244	257	186	443	431	314	220
Free CO_2	18	14	16	22	20	16	22	10	22	10	18	23	18	22
Total CO_2	25.04	24.56	26.56	32.56	28.8	23.04	32.56	20.56	36.8	25.04	44.56	28.8	19.76	2.88
TA	80	120	120.32	121.14	100.16	80.06	100.32	121.63	160	121.96	118.36	101.8	37.55	22.39
TH	184	172	164	152	128	106	88	92	120	84	160	204	94	76
Ca Hard	170.02	141.67	117.3	107.8	108.9	86.25	67.5	71.25	97.7	63.8	145.8	187.6	76.43	62.8
Mg Hard	13.97	30.32	47.7	44.2	20.1	19.75	20.5	20.75	22.3	4.2	14.2	16.4	17.56	13.2
DO	5.3	5.71	4.48	3.67	4.89	4.89	3.67	9.79	3.68	6.12	4.89	5.71	6.73	5.21
Salinity	153.58	204.82	128.02	128.02	140.82	115.22	76.81	115.22	89.63	64.02	153.61	153.61	89.63	64.03
Na	54.63	46.88	38.13	34.88	39.63	39.38	21.31	27.19	26.19	10.78	45.88	39.63	32.82	23.24
K	21.88	20.55	5.79	8.33	8.72	2.63	1.8	17.34	3.01	0.74	10.94	6.62	2.81	2.81
Ca	68.01	56.67	46.92	43.12	43.56	34.5	27	28.5	39.08	25.52	58.32	75.04	30.57	25.12

Sample ID / Parameter	A1	A2	A3	A4	A5	A6	A7	A8	A9	A10	A11	A12	A13	A14
Mg	3.49	7.58	11.67	11.05	7.77	4.93	5.12	5.18	5.57	1.05	3.55	4.1	4.39	3.3
Cl^-	85.09	113.46	70.91	70.91	78	63.82	42.54	63.82	49.64	35.46	85.09	85.09	49.64	35.46
HCO_3^-	97.6	146.4	146.4	146.4	122	97.6	12.2	14.64	19.52	4.88	14.64	12.2	2.44	1.22
NO_3^-	28.15	19.88	34.69	21.23	23.23	16.67	13.59	19.24	13.02	21.38	43.21	37.18	24.47	6.04
SO_4^{2-}	20.81	10.33	7.22	12.65	11.66	7.22	7.22	8.11	9.33	8.11	12.44	9.44	6.22	5.44

Sample ID Parameter	A15	A16	A17	M1	M2	M3	M4	S1	S2	S3	S4	S5	S6	S7
Temp.	14.8	12.8	13.4	14.9	14	13	16.8	20	15	17.9	16.4	15.6	16	15.7
pH	6.27	6.29	6.46	6.47	6.31	6.65	7.2	7.55	7.02	6.46	6.87	6.51	6.62	6.56
Eh	51	49	40	38	69	27	-13	-40	-1	39	12	37	28	28
EC	657	651	224	194	128	224	504	721	159	129	147	248	109	123
TDS	416	414	147	124	81	141	323	460	100	81	93	181	68	78
Free CO_2	26	18	18	18	20	20	4	4	6	24	13	24	16	14
Total CO_2	26.88	18.88	39.252	25.04	27.04	25.28	13.68	19.84	13.04	31.04	31.04	31.04	19.52	31.04
TA	69.47	38.35	24.15	80	80.81	60	110	180	80.81	79.18	100.32	80	40	79.8
TH	148	138	75	44	38	64	154	208	46	46	54	78	20	32
Ca Hard	142.8	132.9	60.93	29.22	25.35	51.8	117.9	188.1	43.5	40.5	46.47	58.53	15.15	27.85
Mg Hard	5.2	5.06	14.07	14.77	12.65	12.2	36.02	19.9	2.5	5.5	18.59	19.96	4.85	4.15
DO	5.71	5.71	9.14	13.87	5.71	6.52	5.71	5.71	5.3	8.16	8.56	6.52	4.89	5.3
Salinity	153.63	115.22	51.23	38.42	25.62	64.01	115.22	134.82	12.82	25.62	12.82	64.02	25.62	25.62
Na	34.27	39.65	6.14	9.09	4.68	1.36	12.96	73.76	8.21	9.28	9.21	9.21	18.81	4.89
K	6.29	8.29	0.81	3.24	1.2	8.9	8.95	2.94	6.93	0.82	0.78	5.93	4.08	3.33
Ca	57.02	53.18	24.41	11.69	10.14	20.72	47.11	75.24	17.4	16.14	18.58	23.41	6.06	11.14
Mg	1.3	1.26	3.43	3.69	3.16	3.05	9	4.97	0.625	1.38	4.64	4.86	1.21	1.03
Cl^-	85.1	63.82	28.36	21.27	14.18	35.45	63.82	74.6	7.09	14.18	7.09	35.46	14.18	14.18
HCO_3^-	84.76	46.79	29.46	97.6	98.66	73.25	134.21	219.63	98.65	96.67	122.4	97.64	48.83	96.62
NO_3^-	41.29	35.23	0.62	8.62	4.17	9.74	4.29	0.04	1.13	2.45	0.12	0.94	1.27	0.37
SO_4^{2-}	8.33	10.33	6.43	6.22	10.33	10.42	14.64	12.66	6.22	8.22	5.11	6.21	8.14	20.33

Table 3.2: Continuied

Sample ID Parameter	S8	M1	M2	P1	P2	P3	P4	P5	P6	P7	J1	J2	J3	J4	J5
Temp.	17.7	13.6	14	15.5	14.1	16.2	16.2	14.8	15.4	15.8	16	14.8	14	12.4	15.3
pH	6.95	6.73	6.68	7.31	6.4	7.36	6.43	6.3	7.39	6.35	6.04	6.3	6.52	6.72	6.96
Eh	3	21	23	-14	45	-25	38	49	-25	45	68	46	38	21	4
EC	71	252	79	97	147	81	122	158	143	93	141	12	97	78	217
TDS	44	159	49	62	92	51	78	101	92	59	90	76	61	49	137
Free CO_2	4	30	8	4	22	3	14	24	3	16	26	18	10	6	10
Total CO_2	7.52	44.08	11.52	7.52	25.52	7.4	19.28	31.04	10.04	19.52	31.28	23.28	15.28	9.52	17.04
TA	43.27	160	38.36	36.72	39.83	50.16	60	80	77.54	44.91	60	64.09	62.42	44.09	77.54
TH	24	36	18	22	36	24	34	40	32	18	30	34	28	20	90
Ca Hard	18.6	30.47	16.25	20.84	19.15	21.99	15.95	33.3	27.5	15.53	22.87	22.5	24.82	15.82	75.25
Mg Hard	5.4	5.53	1.74	1.59	16.85	2.031	18.05	6.7	4.5	2.462	7.13	11.5	3.18	4.18	14.75
DO	11.42	3.26	8.77	12.64	7.74	8.97	7.75	9.79	8.97	7.75	7.34	8.16	8.56	7.75	7.34
Salinity	12.82	25.62	12.82	12.81	12.82	12.82	38.42	25.62	12.82	12.82	31.65	31.65	31.65	12.82	12.82
Na	1.08	8.65	5.42	5.01	7.48	6.89	10.09	9.46	7.98	9.53	8.65	5.31	6.14	4.32	7.64
K	0.2	1.6	1.51	1.35	1.87	0.35	2.31	3.46	2.6	1.77	2.95	1.82	1.43	0.62	3.48
Ca	7.44	12.24	6.53	8.37	7.66	8.83	6.38	13.32	11.04	6.24	9.15	9	9.93	6.33	30.1
Mg	1.35	1.35	0.42	0.28	4.21	0.49	4.4	1.67	1.13	0.61	1.78	2.87	0.79	1.04	3.68
Cl^-	7.09	14.18	7.09	7.09	7.09	7.09	21.27	14.18	7.09	7.09	17.52	17.52	7.09	7.09	7.09
HCO_3^-	52.8	195.2	46.2	44.8	48.6	61.2	73.2	97.6	94.6	54.8	73.2	78.2	76.2	53.8	94.6
NO_3^-	0.23	0.0063	0.75	0.26	1.38	0.009	0.013	0.006	0.006	0.006	1.36	5.67	0.19	0.10	0.10
SO_4^{2-}	7.22	7.44	11.44	7.22	5.14	7.22	7.22	7.22	8.14	10.33	6.22	8.14	11.33	6.42	7.18

Table 3.2: Contd.

Sample ID Parameter	J6	H1	H2	H3	H4	H5	H6	H7	H8	H9	H10
Temp.	1.2	6.2	6.9	6.7	6.4	5.2	1.8	6.1	2.9	4.6	6.3
pH	8.12	5.81	6.04	6.92	6.25	6.6	7.4	6.26	5.97	6.55	6.52
Eh	-75	67	70	5	65	31	-23	51	72	80	35
EC	129	169	285	231	141	140	98	184	161	301	318
TDS	65	101	171	142	91	85	51	115	119	191	211
Free CO_2	5	28	38	6	70	52	7	29	38	15	6
Total CO_2	45.48	66.72	95.2	79.92	92.88	100.4	36.92	81.8	83.76	60.76	64.08
TA	46	44	65	84	26	55	34	60	52	52	66
TH	54	65	102	104	26	57	42	65	78	119	141
Ca Hard	56.95	53.69	60.68	67.42	22.25	45.96	40.29	42.97	70.92	70.92	101.87
Mg Hard	3.05	11.31	41.31	36.58	3.75	11.04	2.29	22.03	7.08	48.07	39.13
DO	13.8	5.1	5.9	7.1	5.1	13.2	13.1	8.5	8.9	4.1	13.1
Salinity	19.23	29.47	70.43	32.03	150.38	32.03	11.55	32.03	25.63	51.23	38.43
Na	6.21	6.14	2.98	4.01	8.21	5.41	6.24	6.48	10.25	5.24	3.46
K	1.02	0.048	0.81	1.82	6.14	1.44	2.91	1.61	4.98	2.06	0.78
Ca	20.41	21.51	24.31	27.01	8.91	18.41	16.14	17.21	28.41	28.42	40.81
Mg	0.74	2.76	10.80	8.93	0.91	2.69	0.55	5.37	1.73	11.73	9.55
Cl^-	10.64	16.31	39	17.73	8.51	17.73	6.38	17.73	14.18	28.36	21.27
HCO_3^-	56.12	53.68	79.3	102.48	31.72	67.1	41.48	73.2	63.44	63.44	80.52
NO_3^-	0.014	2.614	3.014	0.014	0.014	0.014	0.014	0.014	0.1824	0.014	8.141
SO_4^{2-}	4.85	8	6.44	8.33	8.33	20.44	8	13.33	9.33	8.33	7.21

Note: All the values are in mg/L except conductivity (µS/cm); Eh. redox potential (mV); salinity (%); temperature (C) and pH. TA-Total Alkalinity, TH-Total Hardness

Table 3.3: Analytical result of spring/ Hand Pump Water sample in the Almora Tehsil (Mansoon, 2007)

Sample ID Parameter	A1	A2	A3	A4	A5	A6	A7	A8	A9	A10	A11	A12	A13	A14
Temperature	1.7	2.5	5	4.2	5.1	4.4	6.8	7.5	7.4	3.1	1.7	1.9	4.2	7.3
pH	6.7	6.89	6.82	6.62	6.59	6.71	6.67	7.28	6.62	6.53	6.64	6.26	6.33	6.63
Eh	23	9	16	30	32	24	25	-22	27	35	25	41	47	27
EC	726	749	619	546	594	530	317	424	505	341	676	716	471	144
TDS	464	477	391	348	375	337	202	272	319	216	430	452	300	89
Free CO_2	30	30	28	58	46	20	60	6.25	71.25	17.5	28	72	32	30
Total CO_2	89.84	107.44	87.84	114.32	100.56	60.48	98.72	40.35	174.65	43.3	86.08	130.08	60.16	59.92
TA	68	88	68	64	62	46	44	38.75	117.5	30	66	66	32	34
TH	222	216	196	160	174	150	112	116	166	118	206	202	130	46
Ca Hard	169.25	158.55	159.6	127.05	123.9	122.85	74.55	74.55	148.05	84	159.6	149.1	106.05	40.95
Mg Hard	52.75	57.45	36.4	32.95	50.1	27.15	37.45	41.45	17.95	34	46.4	52.9	23.95	5.05
DO	13.6	13.6	10.8	13.4	13.6	13.4	12.2	11.25	10.5	13.75	13.4	9.6	12	13
Salinity	172.82	204.82	147.22	128.02	128.02	153.62	83.23	108.82	102.43	76.83	172.82	192.02	115.22	38.43
Na	49.97	42.32	34.38	29.42	36.22	37.37	17.52	24.75	24.23	7.25	41.75	37.59	29.79	21.13
K	18.65	17.42	3.25	6.62	7.32	1.52	1.24	15.62	2.07	0.54	9.83	5.06	1.72	1.54
Ca	67.7	63.49	63.92	50.88	49.62	49.19	29.85	29.85	59.29	33.64	63.92	59.71	42.47	16.39
Mg	12.87	14.02	8.88	8.04	12.22	6.62	9.14	10.11	4.38	8.29	11.32	12.9	5.84	1.23
Cl^-	95.73	113.46	81.55	70.91	70.91	85.09	46.09	60.28	56.73	42.55	95.73	106.37	63.82	21.27
HCO_3^-	82.96	107.36	82.96	78.08	75.64	56.12	53.68	47.27	143.35	36.6	80.52	80.52	39.04	41.48
NO_3^-	25.13	17.89	33.65	19.56	25.13	17.88	11.58	21.53	11.04	20.37	41.72	38.81	22.46	4.04
SO_4^{2-}	23.33	13.33	8	13.66	12.66	8	8	9.33	10.33	9	13.66	10.33	7.33	6.66

Table 3.3: Contd.

Sample ID Parameter	A15	A16	A17	M1	M2	M3	M4	S1	S2	S3	S4	S5	S6	S7
Temperature	3.4	8.2	4.2	5.9	2.5	4.5	1.5	8.8	10.9	8.5	7.3	7.1	3.7	5.7
pH	6.28	6.44	6.63	6.44	5.73	6.08	6.93	7.42	7.08	6.77	6.83	5.92	5.81	5.94
Eh	50	40	44	41	91	66	5	-32	-3	18	14	77	86	76
EC	652	630	238	184	140	267	468	516	231	137	142	239	162	74
TDS	410	399	152	118	89	169	298	329	147	87	89	152	103	46
Free CO_2	62	58	16.25	38	38	42	9	29	21	44	30	61	37	16
Total CO_2	93.68	109.04	41.55	74.08	74.08	88.64	71.48	144.28	67.64	82.72	76.64	107.64	67.5	43.28
TA	36	58	28.75	41	41	53	71	131	53	44	53	53	35	31
TH	170	162	82	66	57	107	241	125	70	49	52	91	62	37
Ca Hard	139.65	132.3	66.15	67.17	54.58	85.02	198.38	85.02	56.68	47.23	51.43	71.37	47.23	36.74
Mg Hard	30.33	29.7	15.85	ND	2.42	21.98	42.62	39.98	13.32	1.77	0.57	19.63	14.76	0.26
DO	13	13.6	10.75	12.6	13.6	12.8	13.6	13.2	12.8	12.8	12	13	12.6	12.2
Salinity	85.09	81.55	35.46	21.27	28.36	42.55	70.91	24.82	14.18	14.18	8.86	24.82	24.82	7.09
Na	153.62	147.22	64.03	38.43	51.23	76.83	128.02	44.83	25.63	25.63	16.03	44.83	44.83	12.83
K	43.92	70.76	35.07	50.02	50.02	64.66	86.62	159.82	64.66	53.68	64.66	64.66	42.7	37.82
Ca	29.15	36.34	7.25	7.54	3.59	0.96	10.85	69.86	7.53	8.12	8.72	8.89	17.97	3.98
Mg	4.19	6.69	0.94	2.15	0.93	7.84	7.84	2.21	5.14	0.75	0.86	4.86	3.85	2.54
Cl^-	55.93	52.98	26.49	26.91	21.87	34.06	79.48	34.06	22.71	18.92	20.61	28.59	18.92	14.72
HCO_3^-	7.41	7.25	3.87	ND	0.59	5.36	10.39	9.75	3.22	0.43	0.14	4.79	3.60	0.06
NO_3^-	38.56	33.24	0.78	2.72	1.95	4.81	0.85	0.02	1.36	4.94	0.31	4.16	4.5	0.02
SO_4^{2-}	9	11.66	4	11.33	11	10	16.66	13.33	7.33	9.33	7.33	7.66	10.66	25.33

Table 3.3: Contd.

Sample ID Parameter	S8	M1	M2	P1	P2	P3	P4	P5	P6	P7	J1	J2	J3	J4
Temperature	8.9	0.6	6.2	3	2.1	5.6	3.9	4.3	8.6	4.8	12.1	8.5	4.6	1.7
pH	5.99	6.52	5.91	5.83	6.21	7.26	6.53	5.92	6.11	5.47	5.38	5.68	6.75	6.6
Eh	72	35	79	84	57	-16	36	77	61	110	118	94	20	30
EC	146	204	107	78	165	80	124	269	134	94	191	138	119	85
TDS	93	130	68	49	106	50	79	171	85	59	128	87	75	54
Free CO_2	14	29	17	7	66	7	45	11	7	43	53	39	27	25
Total CO_2	54.48	84.44	42.52	29	87.12	29	74.92	28.6	43.08	68.52	82.04	76.84	56.04	51.4
TA	46	63	29	25	24	25	34	20	41	29	33	43	33	30
TH	49	75	38	30	57	27	41	85	50	23	61	56	37	37
Ca Hard	52.48	62.98	34.64	28.34	45.13	28.34	34.64	72.42	43.03	25.02	55.63	48.28	34.64	29.39
Mg Hard	-3.48	12.02	3.36	1.66	11.87	-1.34	6.36	12.57	6.95	-2.02	5.37	7.72	2.36	7.61
DO	13.6	10.6	12.8	13.8	13	13.6	13.2	13.2	14	12.2	12.2	12.3	12.2	13.8
Salinity	19.23	25.63	19.23	12.83	51.23	19.23	38.43	70.43	25.63	12.83	57.63	19.23	25.63	25.63
Na	0.92	7.42	4.61	4.74	5.97	5.89	9.01	8.23	8.75	8.87	7.65	4.98	5.98	3.12
K	0.12	0.98	1.21	1.04	0.89	0.23	1.73	3.07	1.87	1.87	2.14	1.76	1.23	0.43
Ca	21.03	25.23	13.88	11.36	18.08	11.35	13.88	29.01	17.24	10.09	22.29	19.34	13.88	11.77
Mg	-0.85	2.93	0.82	0.41	2.89	-0.33	1.55	3.07	1.69	-0.49	1.31	1.88	0.57	1.86
Cl^-	10.64	14.18	10.64	7.09	28.36	10.64	21.27	39	14.18	7.09	31.91	10.64	14.18	14.18
HCO_3^-	56.12	76.86	35.38	30.5	29.28	30.5	41.48	24.4	50.02	35.38	40.26	52.46	40.26	36.6
NO_3^-	0.04	0.02	2.28	0.02	7.07	0.26	0.63	12.94	0.02	0.02	2.84	0.02	0.02	0.02
SO_4^{2-}	8.66	8.33	12.33	8.33	6	8.33	8.33	8.33	10.67	17.67	7	10.67	12	7.67

Table 3.3: Contd.

Sample ID Parameter	J5	J6	H1	H2	H3	H4	H5	H6	H7	H8	H9	H10
Temperature	2.1	0.6	6.7	7.1	7.1	6.8	5.8	2.4	5.8	2.4	3.9	5.6
pH	6.64	8.03	5.92	6.12	6.96	6	6.58	7.2	6.22	5.77	6.62	6.21
Eh	26	-73	78	62	2	69	30	-14	54	88	27	59
EC	111	124	172	291	247	148	146	103	210	173	293	326
TDS	70	79	110	184	157	96	93	65	134	110	186	206
Free CO_2	42	6	31	40	7	68	56	9	31	43	17	8
Total CO_2	63.12	48.24	68.84	93.68	78.28	94.4	106.16	40.68	85.56	91.4	64.52	63.44
TA	24	48	43	61	81	30	57	36	62	55	54	63
TH	50	56	67	99	96	29	59	44	67	80	111	136
Ca Hard	41.98	53.53	58.78	66.13	72.42	27.29	48.28	43.03	49.33	73.47	76.62	107.06
Mg Hard	8.01	2.47	8.22	32.87	23.57	1.91	10.72	0.95	17.67	6.53	34.38	28.94
DO	12.4	13.2	7.2	6.6	6.8	5.4	13.8	13.6	11.4	8.2	6.2	12.4
Salinity	19.23	12.83	32.03	64.03	25.63	19.23	25.63	12.83	25.63	32.03	64.03	51.23
Na	6.25	0.05	5.67	3.24	4.34	9.87	5.52	5.97	6.59	11.45	5.69	3.78
K	0.32	1.04	0.05	0.79	1.54	6.65	1.74	2.54	1.21	5.32	2.11	0.65
Ca	16.82	21.45	23.55	26.49	29.02	10.93	19.34	17.24	19.76	29.44	30.69	42.89
Mg	1.95	0.60	2.01	8.02	5.75	0.42	2.62	0.23	4.31	1.59	8.39	7.06
Cl^-	10.64	7.09	17.73	35.46	14.18	10.64	14.18	7.09	14.18	17.73	35.46	28.36
HCO_3^-	29.28	58.56	52.46	74.42	98.82	36.6	69.54	43.92	75.64	67.1	65.88	76.86
NO_3^-	0.02	0.02	1.32	3.32	0.02	0.02	0.02	0.02	0.02	0.26	0.02	9.73
SO_4^{2-}	8.33	5	10	7.33	10.67	8	19.33	8	12.67	10.67	9	8.33

Note: All the values are in mg/L except conductivity (µS/cm); Eh. redox potential (mV); salinity (%); temperature (C) and pH. TA-Total Alkalinity, TH-Total Hardness

Table 3.4: Analytical result of Spring/Hand Pump water samples in the Almora Tehsil area (Post-Monsoon, 2007)

Sample ID Parameter	A1	A2	A3	A4	A5	A6	A7	A8	A9	A10	A11	A12	A13	A14
Temperature	-3.4	-3.5	-4	-3	-3.4	-4.7	-4	22.8	-3.1	-0.4	-4.1	-3.7	-2.8	-2.9
pH	6.39	6.96	6.04	5.95	6.34	5.98	5.66	6.58	7.17	6.92	5.77	6.72	5.84	5.65
Eh	42	3	67	63	48	73	97	32	-9	7	87	21	83	97
EC	539	534	515	452	450	414	277	339	386	301	517	563	373	272
TDS	342	340	327	287	286	263	175	214	247	328	358	236	173	321
Free CO_2	12	14	12	22	20	8	26	6	40	16	22	42	10	16
Total CO_2	61.28	87.92	70.08	74.8	69.28	48.48	61.2	50	113.92	37.12	69.52	101.84	29.36	38.88
TA	56	84	66	60	56	46	40	50	84	24	54	68	22	26
TH	222	214	212	188	170	162	114	116	174	120	204	206	126	108
Ca Hard	180.54	159.54	174.24	138.55	149.05	134.35	86.07	100.76	128.05	88.17	161.64	167.94	100.64	10.5.61
Mg Hard	41.46	13.28	9.21	49.45	20.95	27.65	27.93	15.24	45.95	31.83	42.36	38.06	25.36	2.28
DO	12	11.6	12.4	10	10.4	12	10	14	8.8	8.4	13.2	9.2	12.8	12
Salinity	128.03	140.83	140.83	102.43	115.23	128.03	76.83	102.43	102.43	6.43	140.83	153.63	102.43	76.83
Na	51.87	41.02	30.76	27.54	31.71	31.65	19.87	22.32	27.69	10.24	42.23	32.46	25.39	20.87
K	21.98	19.65	5.08	5.68	4.28	5.89	7.67	12.74	6.68	2.75	10.76	6.98	2.82	9.42
Ca	72.33	63.92	69.81	55.51	59.71	53.83	34.48	40.37	51.30	35.32	64.76	67.28	40.37	42.41
Mg	10.12	54.45	37.76	12.06	5.11	6.74	6.82	3.72	11.21	7.77	10.33	9.28	6.18	0.58
Cl^-	70.91	78.00	78.00	56.73	63.82	70.91	42.55	56.73	56.73	3.55	78.00	85.09	56.73	42.55
HCO_3^-	68.32	102.48	8.052	73.2	68.32	56.12	48.8	61	102.48	29.28	65.88	82.96	26.84	31.72
NO_3^-	28.41	19.49	36.94	20.64	28.15	20.68	14.69	23.62	15.03	22.94	45.63	37.54	24.98	5.69
SO_4^{2-}	26.22	14.66	9.33	14.22	13.66	9.33	9.33	10.33	11.21	9.33	14.66	11.42	8.33	7.11

Table 3.4: Contd.

Sample ID Parameter	A15	A16	A17	M1	M2	M3	M4	S1	S2	S3	S4	S5	S6	S7
Temperature	-3.9	-2	2.6	0.8	-2.5	-4.3	0	1	-2	0.8	0.7	0.4	0.3	2.4
pH	6.36	5.88	7.07	5.56	7.2	6.71	7.19	6.56	6.31	5.31	6.47	5.5	5.44	5.72
Eh	47	80	-2	105	-10	15	-11	48	54	120	106	108	113	92
EC	506	516	222	199	134	235	497	395	132	131	153	214	91	78
TDS	321	328	141	126	84	149	314	84	84	253	98	136	58	49
Free CO_2	32	34	24	8	12	28	110	16	6	16	12	14	12	6
Total CO_2	67.2	77.76	54.88	65.44	63.68	61.92	91.84	151.68	58.4	68.96	72.48	68.96	53.12	58.4
TA	40	52	26	38	36	34	68	136	30	42	46	42	24	30
TH	200	182	84	88	60	108	260	124	64	60	64	80	48	48
Ca Hard	170.04	133.24	71.37	67.17	48.28	94.46	195.23	88.17	46.18	52.48	56.68	67.17	27.29	37.78
Mg Hard	7.31	11.89	3.08	20.82	11.72	13.53	64.78	35.83	17.82	7.52	7.32	12.82	20.71	10.21
DO	10.4	10	8.4	10	10	10	8.8	8.4	12.4	10	11.2	13.2	9.6	12.8
Salinity	140.83	140.83	51.23	38.43	12.83	64.03	128.03	25.63	12.83	25.63	12.83	38.39	25.63	12.83
Na	32.42	37.43	9.21	7.98	4.85	2.13	10.62	72.18	8.32	8.21	8.11	7.75	17.98	4.62
K	4.66	7.17	2.63	2.15	1.53	0.67	6.92	2.12	2.12	5.41	0.75	4.24	3.62	1.98
Ca	68.12	53.38	28.59	26.91	19.34	37.85	78.22	35.32	18.83	25.63	22.71	26.91	10.93	15.14
Mg	7.31	11.89	3.08	5.08	2.86	3.30	15.80	8.74	4.35	1.83	1.78	3.13	5.05	2.49
Cl^-	78.00	78.00	28.36	21.27	7.09	35.46	70.91	14.18	7.09	14.18	7.09	21.27	14.18	7.09
HCO_3^-	48.8	63.44	31.72	46.36	43.92	41.48	82.96	165.92	36.6	51.24	56.12	51.24	29.28	36.6
NO_3^-	42.69	35.05	1.07	4.11	3.94	4.21	0.92	0.04	1.69	6.25	0.53	5.25	3.25	1.98
SO_4^{2-}	9.33	12.44	8.11	12.66	11.33	12.66	18.33	14.21	8.33	10.33	8.33	7.82	11.21	28.14

Table 3.4: Contd.

Sample ID Parameter	S8	M1	M2	P1	P2	P3	P4	P5	P6	P7	J1	J2	J3	J4
Temperature	-3.2	-3.5	-2.5	0.8	-3.4	0.8	2	1.2	-0.5	-0.8	-1	-4	-0.6	-4.1
pH	5.73	7.12	5.94	5.2	6.47	6.71	6.36	6.93	6.06	7.11	5.43	6.1	6.65	6.86
Eh	90	-7	77	129	36	23	43	6	69	-4	112	105	25	12
EC	117	275	83	92	123	120	191	84	106	70	133	111	106	78
TDS	74	180	52	58	78	76	121	52	66	49	84	71	67	49
Free CO_2	8	92	6	12	8	10	6	10	8	16	12	16	14	4
Total CO_2	46.72	192.32	25.36	38.4	53.76	41.68	28.88	22.32	32.64	36.36	36.64	56.48	49.2	25.12
TA	44	114	22	30	52	36	26	14	28	22	28	46	40	24
TH	72	110	46	34	62	44	36	66	52	40	46	52	44	40
Ca Hard	48.28	79.77	33.58	31.48	44.08	48.28	23.09	56.68	37.78	31.48	39.88	58.78	35.68	31.48
Mg Hard	23.72	30.23	12.41	2.51	17.92	-4.28	12.91	9.32	14.21	8.51	6.11	-6.78	8.31	8.51
DO	7.6	3.6	10.8	12.8	13.2	10	11.2	10.4	12.8	12	10.4	13.2	9.6	12
Salinity	12.83	25.63	12.83	6.43	12.83	25.63	12.83	38.43	12.83	6.43	25.63	6.39	12.83	6.39
Na	1.15	6.39	5.64	5.03	6.82	5.48	9.02	8.56	6.68	8.32	8.45	5.38	5.65	7.24
K	0.76	1.24	1.23	1.12	1.56	0.43	1.87	2.46	1.87	1.45	2.87	1.23	1.24	0.89
Ca	19.34	31.96	13.46	12.62	17.66	17.34	9.25	22.72	15.14	12.62	15.98	20.54	14.29	12.62
Mg	5.78	7.37	3.03	0.61	4.37	0.19	3.15	2.27	3.47	2.07	1.49	0.20	8.31	8.51
Cl^-	7.09	14.18	7.09	3.54	7.09	14.18	7.09	21.27	7.09	3.54	14.18	3.54	7.09	3.54
HCO_3^-	53.68	139.08	26.84	36.6	63.44	43.92	31.72	17.08	34.16	26.84	34.16	56.12	48.8	29.28
NO_3^-	0.52	0.04	3.39	0.09	8.83	0.54	0.86	10.21	0.07	0.07	3.91	0.09	0.06	0.005
SO_4^{2-}	9.22	10.21	14.61	9.33	7.27	9.22	9.33	9.33	11.66	12.66	7.67	11.33	13.41	7.91

Sample ID Parameter	J5	J6	H1	H2	H3	H4	H5	H6	H7	H8	H9	H10
Temperature	-4.5	-1.7	-2.6	-6.4	2.5	0.05	1.3	-1.8	3.3	-1.2	1.3	2.9
pH	7.33	7.13	7.4	6.52	6.46	6.94	6.21	6.25	6.55	6.6	6.26	5.97
Eh	-24	-7	-23	35	40	6	100	67	81	30	52	73
EC	191	171	212	66	213	163	210	115	209	186	246	276
TDS	122	109	138	41	136	105	133	72	134	119	155	175
Free CO_2	12	10	12	6	10	16	52	6	12	14	10	20
Total CO_2	70.08	52.24	80.64	32.4	105.04	79.36	117.12	37.68	73.6	61.52	71.6	81.6
TA	66	48	78	30	108	72	74	36	70	54	70	70
TH	110	66	112	32	102	74	90	42	102	90	140	138
Ca Hard	79.77	48.28	109.16	33.58	77.67	62.98	90.27	46.18	56.68	71.37	69.27	92.37
Mg Hard	30.23	17.72	2.84	-1.58	24.33	11.02	-0.27	-4.18	45.32	18.62	70.72	45.64
DO	4.8	6.8	6	14	8	8	6	12.8	6	9.6	8.4	10
Salinity	6.39	12.83	25.63	6.39	25.63	12.83	12.83	6.39	6.39	38.43	64.03	64.03
Na	7.48	9.82	7.74	4.93	6.21	12.34	6.25	7.24	8.24	14.24	8.23	4.21
K	1.52	2.54	0.25	1.65	1.92	8.12	2.24	4.45	2.54	6.83	3.54	1.62
Ca	31.95	19.34	43.73	13.45	31.12	25.23	36.16	18.50	22.71	28.59	27.75	37.00
Mg	7.37	4.32	0.69	-0.38	5.93	2.68	-0.06	1.02	11.05	4.54	17.25	11.13
Cl^-	3.54	7.09	14.18	3.54	14.18	7.09	7.09	3.54	3.54	21.27	35.45	35.45
HCO_3^-	80.52	58.56	95.16	36.6	131.76	87.84	9.03	43.92	85.4	65.88	85.4	85.4
NO_3^-	0.05	0.08	2.25	4.45	0.05	0.75	0.85	0.97	0.25	0.73	0.69	10.25
SO_4^{2-}	9.41	6.11	11.66	7.99	11.33	9.223	20.41	9.33	13.66	11.24	9.33	9

Note: All the values are in mg/L except conductivity (µS/cm); Eh. redox potential (mV); salinity (%); temperature (C) and pH. TA-Total Alkalinity, TH-Total Hardness

Table 3.5: Analytical result of Spring/Hand Pump water samples in the Almora Tehsil area (Pre-Monsoon, 2008)

Sample ID Parameter	A1	A2	A3	A4	A5	A6	A7	A8	A9	A10	A11	A12	A13	A14
Temperature	13.2	14.7	13.2	15.5	12.3	16.2	14	16.2	14.5	12.4	13.6	16.3	13.2	15.4
pH	6.24	6.7	6.25	6.39	6.6	6.48	6.58	6.72	6.58	6.39	6.48	6.55	6.58	6.34
Eh	51	22	52	39	31	37	32	21	32	42	37	29	32	48
EC	701	627	610	502	500	428	385	403	282	232	701	640	506	369
TDS	450	430	372	304	315	204	250	209	159	185	450	442	324	230
Free CO_2	18	16	12	24	22	18	24	12	24	12	20	26	18	24
Total CO_2	86.64	130.4	117.6	131.36	111.76	86.64	112	121.12	157.76	122.88	125.6	117.52	47.92	46.88
TA	78	130	120	122	102	78	100	124	152	126	120	104	34	26
TH	182	176	166	150	122	102	80	94	126	90	171	230	96	78
Ca Hard	168	151.2	128.1	149.05	109.2	94.5	73.5	73.5	100.8	73.5	149.1	189	73.5	65.1
Mg Hard	14	24.8	37.9	0.95	12.8	7.5	6.5	20.5	25.2	16.5	21.9	41	22.5	12.9
DO	6.2	5.3	4.62	3.25	4.25	4.25	3.26	10.2	3.25	6.36	3.26	5.98	6.26	6.02
Salinity	159.11	243.22	140.82	140.82	153.62	128.02	89.63	128.02	102.43	76.81	76.81	102.43	76.67	166.42
Na	56.82	49.24	40.81	36.24	40.21	41.42	22.12	28.11	28.01	12.81	48.21	41.21	33.21	24.23
K	25.05	25.08	6.25	9.21	9.81	3.11	2.1	18.15	3.45	1.24	12.81	7.12	3.12	1.01
Ca	67.31	60.58	51.32	59.72	48.88	37.86	29.25	29.25	40.38	29.25	59.74	75.72	29.25	26.08
Mg	3.42	6.05	9.25	0.23	3.12	1.83	1.59	5.00	6.15	4.03	5.34	10	5.49	3.15
Cl^-	88.13	134.73	78	78	85.09	70.91	49.64	70.91	56.73	42.55	42.55	56.73	42.46	92.18
HCO_3^-	95.16	158.6	146.4	148.84	124.44	95.16	122	151.28	185.44	153.72	146.4	126.68	41.48	31.72
NO_3^-	29.12	21.22	36.67	25.12	26.11	20.11	12.21	14.12	21.64	14.69	25.28	45.61	38.21	26.11
SO_4^{2-}	24.61	90.24	8.15	13.41	12.41	9.11	9.11	9.21	10.33	8.46	13.66	8.46	6.99	7.22

Table 3.5: Contd.

Sample ID Parameter	A15	A16	A17	M1	M2	M3	M4	S1	S2	S3	S4	S5	S6	S7
Temperature	13.2	13.2	16.8	15.6	14.9	13	18	21	17.5	18	17	12	17	16.2
pH	6.29	6.27	7.17	6.46	6.47	6.51	7.17	7.2	7.91	6.39	6.71	6.56	6.47	6.27
Eh	49	51	-9	39	38	37	-9	-10	-11	42	15	48	106	51
EC	647	629	705	183	115	305	545	762	150	75	99	186	69	82
TDS	418	412	455	122	79	150	333	470	99	26	12	23	18	19
Free CO_2	28	26	26	16	22	22	6	6	8	22	15	26	18	16
Total CO_2	93.12	62.96	73.52	88.16	92.4	73.04	111.6	171.44	79.28	92.4	104.76	87.16	33.84	88.16
TA	74	42	54	82	80	58	120	188	81	80	102	82	46	82
TH	144	130	68	46	35	70	162	210	48	47	56	76	24	36
Ca Hard	127.5	124.4	50.08	31.06	28.01	46.18	134.35	175	35.68	39.8	37.79	48.28	14.8	31.49
Mg Hard	16.5	5.6	17.92	14.94	6.99	23.82	27.65	35	12.32	7.2	18.21	27.72	9.2	4.51
DO	5.21	5.61	4.02	13.05	6.21	6.38	5.91	6.02	6.2	9.18	9.39	8.02	5.69	6.2
Salinity	166.42	128.02	76.83	51.228	38.43	76.83	128.02	140.82	25.62	25.62	38.43	128.02	51.23	51.23
Na	35.11	40.21	10.11	9.11	5.21	1.49	13.21	75.11	9.11	10.11	9.81	9.81	19.25	5.25
K	7.18	9.21	2.16	4.11	1.8	9.1	9.21	3.12	7.21	1.21	0.89	6.11	5.06	3.39
Ca	51.08	49.84	20.06	12.44	11.22	18.51	53.83	70.11	14.29	15.95	15.14	19.34	5.93	12.62
Mg	4.03	1.37	4.37	3.65	1.71	5.81	6.75	8.54	3.01	1.76	4.42	6.76	2.24	1.1
Cl^-	92.18	70.91	42.55	28.37	21.27	42.55	70.91	78	14.18	14.18	21.28	70.91	28.37	28.37
HCO_3^-	90.28	51.24	65.88	100.04	97.6	70.76	146.4	229.36	98.82	97.6	124.44	100.04	56.12	100.04
NO_3^-	7.91	31.16	2.05	7.32	3.19	10.15	6.21	0.091	2.15	3.08	0.211	0.981	1.378	0.415
SO_4^{2-}	10.46	7.98	7.48	7.66	11.44	11.21	16.98	13.81	7.44	9.33	6.21	7.34	9.16	23.61

Table 3.5: Contd.

Sample ID Parameter	S8	M1	M2	P1	P2	P3	P4	P5	P6	P7	J1	J2	J3	J4	J5
Temperature	18.2	14.5	13.68	16.2	16.2	14.8	15.9	15.6	15.3	16	17	15.2	13.8	13.3	16.2
pH	5.84	7.36	5.66	6.1	7.36	6.3	6.4	6.35	7.31	6.3	6.31	6.52	6.3	6.43	6.72
Eh	83	-25	97	105	-25	49	45	45	-14	46	54	37	46	38	21
EC	63	298	71	99	153	91	146	153	163	98	146	110	92	87	304
TDS	42	265	44	65	98	56	79	120	98	62	95	79	63	56	153
Free CO_2	6	32	10	6	24	5	16	26	5	18	28	18	8	8	12
Total CO_2	46.48	188.64	41.68	39.44	60.96	52.52	72.32	101.68	73.64	60.24	84.32	70.8	66.08	50.24	80.64
TA	46	178	36	38	42	54	64	86	78	48	64	60	66	48	78
TH	22	38	16	24	38	26	36	44	36	20	32	36	26	22	98
Ca Hard	20.01	31.02	14.63	18.12	30.25	22.01	34.69	38.09	32.08	14.89	28.13	30.18	23.51	14.32	87.16
Mg Hard	1.99	6.98	1.37	5.88	6.75	3.99	1.31	5.91	3.92	5.11	3.87	5.82	2.49	7.68	10.84
DO	11.21	3.92	9.88	13.78	8.23	10.25	9.31	10.05	9.32	8.76	6.21	9.05	6.99	8.01	10.84
Salinity	25.62	38.42	25.62	25.62	25.3	25.3	51.23	38.43	25.63	25.63	32.01	32.01	25.63	25.63	14.18
Na	1.25	9.15	6.11	5.11	8.11	7.25	11.15	10.25	8.15	10.15	9.15	6.21	6.2	5.61	95.16
K	0.8	1.8	1.81	1.41	2.21	0.82	2.81	3.92	3.1	2.12	3.1	2.11	2.55	1.25	8.11
Ca	8.01	12.43	5.86	7.259	12.12	8.82	13.89	15.26	12.85	5.97	11.27	12.09	9.41	5.74	34.92
Mg	0.485	1.703	0.334	1.434	1.647	0.973	0.319	1.442	0.956	1.246	0.944	1.42	0.6075	1.873	2.644
Cl^-	14.18	21.27	14.18	14.18	14	14	28.37	21.27	14.18	14.18	17.72	17.72	14.18	14.18	7.75
HCO_3^-	56.12	217.16	43.92	46.36	51.24	65.88	78.08	104.92	95.16	58.56	78.08	73.2	80.52	58.56	25.63
NO_3^-	0.371	0.012	0.812	0.321	1.51	0.012	0.021	0.01	0.02	0.01	1.751	4.281	0.215	0.121	0.121
SO_4^{2-}	8.41	8.41	12.61	8.33	6.21	6.33	6.33	6.33	9.33	11.41	7.25	9.33	11.66	7.45	7.91

Table 3.5: Contd.

Sample ID Parameter	J6	H1	H2	H3	H4	H5	H6	H7	H8	H9	H10
Temperature	12.5	13.2	15	16.7	18.9	19.6	20.1	17.6	16.8	15.3	14.8
pH	7.17	7.11	6.93	6.47	6.86	6.1	6.21	6.55	6.26	6.21	5.95
Eh	-9	-4	6	36	12	105	100	80	52	100	63
EC	191	203	120	217	152	141	92	126	179	303	283
TDS	122	143	76	135	90	93	7	15	125	160	180
Free CO_2	12	10	8	8	18	48	8	14	10	12	22
Total CO_2	45.44	83.92	37.68	113.6	84.88	116.64	43.2	82.64	64.56	78.88	88.88
TA	38	84	36	120	76	78	40	78	62	76	76
TH	60	102	36	112	80	92	46	108	56	74	74
Ca Hard	54.21	91.08	31.05	98.11	74.08	79.01	41.04	95.07	44.09	69.14	68.17
Mg Hard	5.79	10.92	4.95	13.89	5.92	12.99	4.96	12.93	11.91	4.86	5.83
DO	8.06	9.21	8.31	8.23	6.58	3.29	6.05	10.62	3.25	6.39	9.24
Salinity	25.63	25.63	38.43	32.02	12.8	19.23	12.8	12.8	19.23	51.23	38.43
Na	8.01	9.89	8.12	5.11	6.81	13.11	7.15	8.11	15.11	9.15	4.81
K	1.89	3.11	1.31	1.51	2.21	9.1	2.12	3.11	7.12	4.21	1.91
Ca	21.72	36.49	12.44	39.31	29.68	31.65	16.44	38.09	17.66	27.7	27.31
Mg	1.41	2.66	1.21	3.39	1.44	3.17	1.21	3.15	2.97	1.18	1.42
Cl^-	14.18	14.18	21.27	17.73	7.0914	10.63	7.091	7.091	10.64	28.37	21.27
HCO_3^-	46.36	102.48	43.92	146.4	92.72	95.16	48.8	95.16	75.64	92.72	92.72
NO_3^-	0.02	2.38	4.82	0.06	0.81	0.91	0.99	0.31	0.81	0.71	11.35
SO_4^{2-}	5.91	8.33	7.12	9.33	9.33	23.41	8.33	14.66	8.41	7.61	7.91

Note: All the values are in mg/L except conductivity (µS/cm); Eh. redox potential (mV); salinity (%); temperature (C) and pH. TA-Total Alkalinity, TH-Total Hardness

Table 3.6: Analytical result of Spring/Hand Pump water samples in the Almora Tehsil area (Monsoon, 2008)

Sample ID Parameter	A1	A2	A3	A4	A5	A6	A7	A8	A9	A10	A11	A12	A13	A14
Temperature	1.9	2.8	6	4.1	5.3	5.1	6.9	7.8	7.5	4.1	1.9	2.1	5.1	8.1
pH	6.71	6.82	6.81	6.67	6.62	6.89	6.63	7.02	6.46	6.51	6.65	6.31	6.46	6.65
Eh	24	16	9	25	30	9	27	-1	39	37	27	69	39	27
EC	711	732	625	521	589	528	325	450	515	346	689	752	485	154
TDS	452	481	410	352	381	341	215	274	325	210	462	459	298	91
Free CO_2	32	32	32	62	48	18	64	6.35	71.12	18.2	30	74	30	34
Total CO_2	109.31	124.33	102.86	132.85	116.71	69.53	113.38	49.294	199.95	54.7	103	147	66.5	68.36
TA	72	86	66	66	64	48	46	40	120	34	68	68	34	32
TH	226	218	206	162	178	154	116	118	168	120	240	206	136	48
Ca Hard	165.9	159.62	160.2	130.2	128.1	121.8	74.8	74.2	147	86.1	157.5	151.2	105	46.2
Mg Hard	60.1	58.38	45.8	31.8	49.9	32.2	41.2	43.8	121	33.9	82.5	54.8	31	1.8
DO	13.5	13.2	13.5	13.2	12.2	12.1	13.1	12.15	12.1	13.73	12.8	12.1	12	13
Salinity	166.42	198.42	153.61	140.82	140.82	160.03	89.63	115.23	108.84	83.22	166.42	198.43	121.61	44.83
Na	50.82	44.61	34.32	28.15	38.15	38.15	18.62	25.12	25.21	7.81	45.15	38.15	34.15	23.15
K	20.62	14.21	2.61	6.65	8.13	1.82	1.82	17.81	2.51	0.74	10.12	6.12	1.91	1.98
Ca	66.47	63.95	64.18	52.16	51.32	48.79	29.97	29.72	58.89	34.49	63.1	60.58	42.06	18.51
Mg	14.66	14.24	11.17	7.76	12.17	7.86	10.05	10.68	29.52	8.27	20.13	13.37	7.56	0.44
Cl^-	92.18	109.91	85.09	78	78	88.64	49.64	63.82	60.28	46.09	92.18	109.92	67.36	24.82
HCO_3^-	87.84	104.92	80.52	80.52	78.08	58.56	56.12	48.8	146.4	41.48	82.96	82.96	41.48	39.04
NO_3^-	27.15	19.92	36.62	20.15	27.124	18.82	12.69	24.21	12.61	22.35	42.73	39.82	24.65	4.04
SO_4^{2-}	25.44	12.66	9	12.66	13.66	9	9	10.33	9.33	9.33	12.66	12.66	8	7.44

Table 3.6: COntd.

Sample ID Parameter	A15	A16	A17	M1	M2	M3	M4	S1	S2	S3	S4	S5	S6	S7
Temperature	3.8	9.1	5.1	5.8	2.8	5.1	1.8	8.1	11.2	9.1	7.8	8.5	4.1	6.5
pH	6.39	6.34	6.58	6.43	5.84	6.04	6.92	7.11	6.93	6.71	6.44	5.73	5.81	5.73
Eh	42	48	32	38	83	67	7	-4	6	15	40	91	86	91
EC	642	615	258	191	138	281	471	510	232	134	144	242	163	82
TDS	391	451	165	119	91	172	314	325	139	82	81	169	151	51
Free CO_2	60	62	18.15	36	36	44	11	32	20	46	34	64	39	14
Total CO_2	102.94	126.42	51.432	81.09	81.09	102.27	94.74	180.15	77.97	95.39	94.12	124.12	79.79	48.36
TA	40	60	31	42	42	58	78	138	54	46	56	56	38	32
TH	174	168	86	68	59	109	244	134	72	48	56	92	64	39
Ca Hard	138.6	130.2	69.3	65.1	56.7	87.75	197.4	84	54.6	46.2	52.5	73.5	46.2	37.8
Mg Hard	35.4	37.8	16.7	2.9	2.3	21.25	46.6	50	17.4	1.8	3.5	18.5	17.8	1.2
DO	13	13.5	12.12	12.5	13.2	12.1	13.1	13.5	12.9	12.2	12	13	12	12
Salinity	160.02	151.07	70.43	44.83	57.63	83.22	140.82	47.39	32.03	32.03	19.23	47.39	47.39	19.23
Na	30.12	38.31	6.15	7.61	3.61	0.97	11.15	70.15	7.61	9.15	9.41	9.81	18.15	4.15
K	5.18	7.12	0.98	3.18	0.81	8.12	8.16	2.15	4.14	0.61	0.92	4.92	4.11	2.56
Ca	55.53	52.16	27.76	26.08	22.72	35.15	79.08	33.65	21.88	18.51	21.03	29.45	18.51	15.14
Mg	8.64	9.22	4.07	0.71	0.56	5.18	11.37	12.2	4.24	0.44	0.85	4.51	4.34	0.29
Cl^-	88.64	83.68	39	24.82	31.91	46.09	78	26.24	17.73	17.73	10.64	26.24	26.24	10.64
HCO_3^-	48.8	73.2	37.82	51.24	51.24	70.76	95.16	168.36	65.88	56.12	68.32	68.32	46.36	39.04
NO_3^-	40.15	34.21	0.77	2.31	2.12	4.81	0.85	0.0012	1.37	5.01	0.32	4.16	4.7	0.05
SO_4^{2-}	9.33	12.33	7.33	12.66	11.33	9.77	18.22	13.33	7.33	9	8	8	12.66	26.22

Table 3.6: Contd.

Sample ID Parameter	S8	M1	M2	P1	P2	P3	P4	P5	P6	P7	J1	J2	J3	J4	J5
Temperature	9.1	0.8	6.3	3.6	3.1	6.2	4.1	4.5	8.7	4.9	13.1	9.1	4.7	2.5	3.1
pH	5.83	6.47	5.92	5.43	6.53	6.75	6.21	6.75	6.1	7.26	5.47	6.21	6.75	6.93	7.26
Eh	84	36	77	112	36	20	-16	20	105	-16	110	57	20	6	-16
EC	149	213	115	81	171	78	131	270	154	101	205	148	112	91	121
TDS	95	150	71	52	115	52	82	181	91	61	129	91	82	61	69
Free CO_2	16	30	19	9	64	9	48	13	9	45	56	41	24	28	46
Total CO_2	67.53	98.71	53.36	37.99	92.79	39.06	86.65	38.77	58.39	78.28	94.65	92.53	62.65	60.21	67.47
TA	48	64	32	27	26	28	36	24	46	31	36	48	36	30	20
TH	51	76	39	32	59	29	42	86	54	26	64	58	38	38	52
Ca Hard	48.3	64.05	33.6	25.2	44.1	26.25	35.7	73.5	44.1	23.1	54.6	46.2	33.6	30.45	44.1
Mg Hard	2.7	11.95	5.4	6.8	14.9	2.75	6.3	12.5	9.9	2.9	9.4	11.8	4.7	7.55	7.9
DO	13	12	13	13	13	13	13	12	11	12	12	12	13	13	13
Salinity	19.24	32.03	12.83	12.83	57.63	12.83	44.83	76.81	32.03	12.83	64.03	19.24	32.03	32.03	12.83
Na	1.21	8.12	5.15	5.1	6.81	6.95	9.15	8.15	9.12	9.06	8.01	5.05	6.15	3.62	7.12
K	0.04	1.01	1.61	1.01	1.25	1.12	1.21	3.08	1.25	2.15	2.18	1.89	1.82	0.56	0.61
Ca	19.35	25.66	13.46	10.09	17.67	10.52	14.3	29.45	17.67	9.25	21.88	18.51	13.42	12.19	17.67
Mg	0.66	2.92	1.31	1.66	3.64	0.67	1.54	3.05	2.42	0.71	2.29	2.87	1.15	1.84	1.93
Cl^-	10.64	17.73	7.09	7.09	31.91	7.09	24.82	42.54	17.73	7.09	35.46	10.64	17.73	17.73	7.09
HCO_3^-	58.56	78.08	39.04	32.94	31.72	34.16	43.92	29.28	56.12	37.82	43.92	58.56	43.92	36.6	24.4
NO_3^-	0.08	0.02	2.31	0.02	7.01	0.31	0.81	13.01	0.05	0.02	2.91	0.005	0.04	0.006	0.01
SO_4^{2-}	9	9	11.33	9.22	6.33	8.33	8.33	8.33	11.22	11.22	10.33	10.72	8.33	5.33	10.33

Table 3.6: COntd.

Sample ID / Parameter	J6	H1	H2	H3	H4	H5	H6	H7	H8	H9	H10
Temperature	0.9	7.1	6.7	6.2	6.9	5.9	2.8	6.1	3.1	4.1	6.1
pH	7.2	7.33	6.55	6.25	6.21	6.52	6.22	6.64	5.97	6.6	6.33
Eh	-14	-24	81	67	59	35	54	26	72	30	52
EC	131	181	301	241	152	148	104	315	181	301	324
TDS	85	151	191	162	81	21	71	141	121	192	215
Free CO_2	8	34	44	9	64	59	12	36	48	16	10
Total CO_2	70.27	83.38	112.7	99.18	100.5	114.83	48.5	106.86	108.12	72.9	83
TA	58	46	64	84	34	52	34	66	56	53	68
TH	58	69	105	98	32	62	46	68	82	115	142
Ca Hard	52.5	60.9	67.2	73.5	29.4	46.2	44.1	48.3	74.34	75.39	109.2
Mg Hard	5.5	8.1	37.8	24.5	2.6	15.8	1.9	19.7	7.66	39.61	32.8
DO	11	8.5	6.8	6.6	5.2	13.2	13.1	12.4	6.6	4.1	11.2
Salinity	33.31	70.43	32.03	20.52	32.03	15.39	33.31	33.31	28.18	70.21	57.63
Na	7.91	6.15	3.62	4.69	10.15	6.61	6.15	7.05	12.61	6.11	3.8
K	1.54	0.06	1.81	1.5	7.21	1.82	3.01	1.31	5.42	2.54	1.05
Ca	21.03	24.39	26.92	29.45	11.78	18.51	17.67	19.35	29.78	30.2	43.75
Mg	1.342	1.97	9.22	5.98	0.63	3.85	0.46	4.81	1.87	9.66	8.003
Cl^-	18.44	39	17.73	11.35	17.73	8.51	18.44	18.44	15.6	39	31.91
HCO_3^-	70.76	56.12	78.08	102.48	41.48	63.44	41.48	80.52	68.32	64.66	82.96
NO_3^-	0.01	0.02	3.615	0.001	0.002	0.004	0.0002	0.21	0.26	0.001	10.21
SO_4^{2-}	8	10.33	8	10.33	8.33	19.34	8.44	13.33	10.33	9.33	8

Note: All the values are in mg/L except conductivity (µS/cm); Eh. redox potential (mV); salinity (‰); temperature (C) and pH. TA-Total Alkalinity, TH-Total Hardness

Table 3.7: Analytical result of Spring/Hand Pump water samples in the Almora Tehsil area (Post-Monsoon, 2008)

Sample ID Parameter	A1	A2	A3	A4	A5	A6	A7	A8	A9	A10	A11	A12	A13	A14
Temperature	-3.8	-3.4	-5.1	-3.6	-4.5	-4.6	-3.8	20.2	-3.9	-0.6	-4.6	-3.9	-2.6	-0.31
pH	6.25	6.94	6.08	5.92	6.44	5.94	5.68	6.6	7.24	6.89	5.74	6.77	5.8	5.7
Eh	52	4	65	76	40	76	93	32	-18	10	92	16	85	91
EC	521	514	510	450	400	412	207	314	392	315	521	541	381	281
TDS	369	310	315	291	271	251	164	204	251	315	361	214	141	301
Free CO_2	10	16	14	20	22	12	28	8	44	18	24	46	12	18
Total CO_2	207.12	91.68	73.84	76.32	73.04	54.24	64.96	53.76	119.68	40.88	73.28	104.08	33.12	42.64
TA	58	86	68	64	58	48	42	52	86	26	56	66	24	28
TH	224	218	218	197	172	171	118	118	176	122	214	218	131	114
Ca Hard	175.41	161.12	176.12	140.51	150.06	138.15	88.05	102.81	130.04	89.12	165.12	162.12	108.61	108.61
Mg Hard	48.59	56.88	41.88	56.49	21.94	32.85	29.95	15.19	45.96	32.88	48.88	55.88	22.39	5.38
DO	11	12	12	13	11	11	12	11	10	10	11	10	11	10
Salinity	134.43	147.23	147.23	108.84	121.63	134.43	83.22	108.84	108.84	12.823	34.41	160.03	108.84	83.57
Na	52.89	42.15	231.15	28.51	32.7	34.15	21.15	24.12	28.14	11.14	43.11	33.51	26.41	21.88
K	23.71	20.15	6.07	6.11	6.15	6.19	8.19	12.14	7.14	3.12	11.23	7.15	3.81	10.18
Ca	70.28	64.55	70.56	56.91	60.12	55.35	35.28	41.19	52.09	35.71	66.15	64.95	43.52	43.52
Mg	11.86	13.8	10.22	13.78	5.35	8.02	7.31	3.71	8.77	8.02	11.93	13.63	5.46	1.31
Cl^-	74.46	81.55	81.55	60.28	67.37	74.46	46.09	60.28	60.28	7.09	74.45	88.64	60.28	46.28
HCO_3^-	70.76	104.92	82.96	78.08	70.76	58.56	51.24	63.44	104.92	31.72	68.32	80.52	29.28	34.16
NO_3^-	29.05	20.15	38.51	22.15	29.16	21.64	15.91	24.63	15.41	23.81	46.14	38.12	25.11	6.1
SO_4^{2-}	24.21	14.41	8.33	14.41	11.44	8.33	8.33	10.44	11.61	9.33	14.64	11.64	8.11	7.21

Table 3.7: Contd.

Sample ID Parameter	A15	A16	A17	M1	M2	M3	M4	S1	S2	S3	S4	S5	S6	S7
Temperature	-3.4	-2.8	3.7	0.9	-3.5	-4.8	0.8	1.6	-2.4	0.9	0.8	0.6	0.3	0.9
pH	6.34	5.8	7.09	5.57	7.28	6.7	7.2	6.7	6.28	5.38	6.53	5.7	5.47	5.83
Eh	46	84	-3	112	-22	20	-18	49	49	120	38	94	111	86
EC	504	501	212	190	139	361	501	381	131	141	159	241	85	81
TDS	302	314	145	128	91	151	324	81	841	265	97	141	41	31
Free CO_2	34	32	26	10	14	30	120	18	8	18	14	16	10	8
Total CO_2	72.72	81.28	50.64	46.96	43.92	61.68	182.48	139.44	32.64	56.72	56.24	54.72	32.88	36.16
TA	44	56	28	42	34	36	71	138	28	44	48	44	26	32
TH	212	181	86	86	62	110	265	126	66	62	62	82	46	50
Ca Hard	172.01	135.31	72.15	69.17	50.29	98.41	200.21	89.17	48.21	54.42	58.11	68.14	29.25	39.15
Mg Hard	39.99	45.69	13.85	16.83	11.71	11.59	64.79	36.83	17.79	7.58	3.89	13.86	16.75	10.85
DO	11	11	9	12	12	12	9	9	11	10.5	12	11	10.2	13.5
Salinity	147.23	147.23	38.43	44.83	19.23	70.43	140.82	32.03	19.23	32.03	19.23	44.83	44.83	19.23
Na	33.82	38.14	10.15	8.15	5.41	2.54	11.71	73.21	9.35	9.41	8.15	7.81	18.15	5.15
K	5.15	8.14	3.71	3.14	169	0.89	7.81	2.81	2.54	6.41	0.81	4.36	3.98	2.54
Ca	68.92	54.21	28.91	27.71	20.15	39.43	80.21	35.73	19.31	21.8	23.28	27.3	11.72	15.68
Mg	9.76	11.15	3.38	4.11	2.86	2.8	15.81	8.98	4.34	1.85	0.95	3.38	4.09	2.65
Cl^-	81.55	81.55	21.27	24.82	10.637	39	78	17.73	10.64	17.73	10.64	24.82	140.82	10.64
HCO_3^-	53.68	68.32	34.16	51.24	41.48	43.92	86.62	168.36	34.16	53.68	58.56	53.68	31.72	39.04
NO_3^-	43.05	35.94	1.08	4.51	4.1	4.61	1.81	0.05	2.15	6.41	0.65	5.61	3.61	2.05
SO_4^{2-}	9.33	12.41	8.21	12.44	11.41	18.41	14.51	8.21	10.41	8.41	7.98	11.21	28.91	10.44

Table 3.7: contd.

Sample ID Parameter	S8	M1	M2	P1	P2	P3	P4	P5	P6	P7	J1	J2	J3	J4	J5
Temperature	-3.8	-3.9	-2.7	1.5	-3.8	0.9	2.5	1.4	-0.54	-0.9	-1.4	-4.5	-0.9	-4.8	-5.4
pH	5.7	7.2	5.91	5.3	6.5	6.75	6.2	7.04	6.1	7.19	5.48	6.11	6.53	6.75	7.4
Eh	94	-18	79	116	34	21	59	-4	62	-16	118	61	36	24	-24
EC	115	281	81	90	115	151	201	81	116	90	136	151	114	81	201
TDS	61	160	41	51	81	81	131	41	69	51	81	80	41	51	131
Free CO_2	10	94	8	14	10	12	6	12	10	10	18	18	16	6	14
Total CO_2	50.48	197.84	29.12	42.16	59.28	45.44	30.64	26.08	36.4	31.12	40.88	42.64	58.24	28.88	73.84
TA	46	118	24	32	56	38	28	16	30	24	26	38	48	26	68
TH	74	112	48	36	68	46	38	68	56	46	44	58	56	42	112
Ca Hard	50.214	82.12	34.51	32.51	46.01	40.12	26.01	57.12	39.712	36.51	40.81	54.121	38.21	32.41	80.15
Mg Hard	23.786	29.88	13.49	3.49	21.99	5.88	11.99	10.88	16.28	9.49	3.91	3.88	17.79	9.59	31.85
DO	8.1	4	10.4	13	11.1	10	12	12	11	13	8	10	12	10	12
Salinity	19.23	32.03	19.23	7.719	19.23	32.03	19.23	44.83	19.23	7.719	32.03	7.719	19.23	7.719	7.719
Na	1.21	7.14	5.81	5.05	7.85	6.14	9.11	9.15	6.91	8.35	8.51	5.41	5.37	7.54	7.91
K	0.85	1.31	1.12	1.81	1.65	1.15	1.91	2.15	1.85	1.54	3.81	1.51	1.54	0.95	1.71
Ca	20.12	32.9	13.83	13.02	18.43	16.07	10.42	22.88	15.91	14.63	16.35	21.68	15.31	12.98	32.11
Mg	5.81	7.29	3.29	0.851	5.36	1.43	2.93	2.65	3.91	2.32	0.95	0.95	4.34	1.36	7.77
Cl^-	10.64	17.73	10.64	4.26	10.64	17.73	10.64	24.82	10.64	4.26	17.73	4.26	10.64	4.26	4.26
HCO_3^-	56.12	143.96	29.28	39.04	68.32	46.36	34.16	19.52	36.6	29.28	31.72	34.16	58.56	31.72	82.96
NO_3^-	0.62	0.06	3.41	0.095	9.85	0.69	0.91	11.25	0.071	0.071	4.51	0.091	0.071	0.004	0.051
SO_4^{2-}	10.44	10.44	11.44	9.14	7.12	9.41	9.41	9.41	11.69	11.69	7.47	11.41	13.66	8.41	8.94

Table 3.7: Contd.

Sample ID Parameter	J6	H1	H2	H3	H4	H5	H6	H7	H8	H9	H10
Temperature	-1.8	-3.4	-7.1	3.1	0.01	1.8	-1.9	3.6	-1.8	1.4	3.1
pH	7.21	7.3	6.47	6.51	6.91	6	6.2	6.6	6.5	6.24	5.9
Eh	-18	-24	106	34	7	69	54	28	80	51	79
EC	182	219	71	241	171	241	121	214	195	245	278
TDS	115	142	51	141	115	135	81	137	210	157	181
Free CO_2	12	14	8	12	18	56	8	14	18	12	22
Total CO_2	56	80.88	39.68	110.56	84.88	124.64	41.44	77.36	67.28	75.36	85.36
TA	50	76	36	112	76	78	38	72	56	72	72
TH	68	116	36	104	78	94	46	106	96	144	140
Ca Hard	46.25	115.17	35.01	79.68	64.98	89.51	45.01	58.21	78.05	71.215	105.05
Mg Hard	21.75	0.83	0.99	24.32	13.02	4.49	0.99	47.79	17.95	72.785	34.95
DO	5	7	7.1	12	9	9	7.4	10	7.3	9.4	10
Salinity	19.23	32.03	7.719	32.03	19.23	19.23	7.719	7.719	44.83	70.43	70.43
Na	9.98	7.81	5.06	6.51	13.15	6.51	7.65	9.15	15.19	9.01	4.51
K	2.61	0.81	1.71	2.04	8.51	2.51	4.51	3.15	7.81	3.81	19.2
Ca	18.53	46.14	14.03	31.92	26.033	35.86	18.03	23.32	31.27	28.53	42.09
Mg	5.307	0.2	0.24	5.93	3.18	1.095	0.2415	11.66	4.38	17.76	8.53
Cl^-	10.64	17.73	4.26	17.73	10.64	10.64	4.26	4.26	24.82	39	39
HCO_3^-	61	92.72	43.92	136.64	92.72	95.16	46.36	87.84	68.32	87.84	87.84
NO_3^-	0.085	2.41	4.61	0.061	0.85	0.81	1.04	0.36	0.81	0.071	11.31
SO_4^{2-}	6.04	9.44	8.41	11.69	8.22	20.33	8.41	16.41	12.11	9.04	9.33

Note: All the values are in mg/L except conductivity (μS/cm); Eh. redox potential (mV); salinity (‰); temperature (C) and pH. TA-Total Alkalinity, TH-Total Hardness

References

Kumar K, Rawat DS and Joshi R, Chemistry of spring water in Almora, central Himalaya, India. *Environ. Geology*, 31(3/4), 150-156 DOI: 10.1007/s002540050174, 1997.

Mishra RC and Sharma RP, Structure of the Almora crystalline, lesser Kumaun Himalayan: An interpretation. *Himalayan Geology*, 2, 330-341, 1972.

Rao PN and Saxena SP, Does Almora Nappe exist? Himalayan Geology, 5, 169-184, 1975.

Desai SJ, Mode of origin and tectonic setting of Gneissic rocks of Siahi Devi area, District Almora, U.P. *Himalayan Geology*, 3, 345-356, 1973.

Pandey RK, Morphometric personally of Bhilangana network (UP Himalaya). *J. Landscape Syst. Ecol.*,8, 34-40, 1985.

Pandey RK and Rawat DS, Morphometric analysis of crystalline and granite rocks in Almora and its environs, In: Shah NK, Bhat SD, and Pande RK (Eds), Himalaya: Environment, Resources and Development, Shree Almora Book Depot, Almora, 102-110.

Srivastava RAK and Gaur Girish CS, Mineralogy and Genesis of sulphide mineralization from Shiskhani-Chhanapani area, district Almora, Kumaun Himalaya. *Himalayan Geology*, **9(2)**, 810-824, 1979.

Agrawal L, Pandey AR and Powar KB, Petrogenesis of Granitic rocks of Almora crystalline mass, Kumaun Himalaya. *Himalayan Geology*, 2, 145-167, 1972.

4

Hydrogeochemical Characterization and Classification of Hill Springs

This chapter deals with hydrogeochemical modeling calculations by using Aquachem4.0 software package. The concentration of major ions usually follow the trend $HCO_3^- > Cl^- > Ca^{2+} > Na^+ > SO_4^{2-} > NO_3^- > Mg^{2+} > K^+$. In springs, ion exchange process with rock composition is suggested by 1.36 equivalent ratio of $Ca^{2+} + Mg^{2+}$ to $Na^+ + K^+$. The 1:1 ratio relationship relates the dissolution of gypsum, anhydrite, calcite and dolomite that controls the major ion chemistry of the springs. Hydrochemical facies of the spring of the study area are characterized by Ca-Na-Cl-HCO_3, Ca-Na-HCO_3-Cl and Ca-HCO_3-Cl water type.

4.1 Hydrogeochemistry

Hydrogeochemistry seeks to determine the origin of the chemical composition of ground water and the relationship between water and rock chemistry, particularly as they relate to ground-water movement. The water chemistry changes as it passes from one geochemical environment to another. The main geological and hydrological factors which generally affect the groundwater geochemistry include rainfall, climate, soil, air, aquifer lithology, saline water and flow pattern. The investigation/ assessment of geochemical element distribution and the natural background of these resources are of paramount importance in reconciling the exploitation of surface and ground waters with the protection of the environment, including the well being of both mankind and local fauna and flora. Because, the chemical (quality) of groundwater is not only related to the lithology of the area and the residence time the water is in contact with rock material, but also reflects inputs from the atmosphere, from soil and weathered mantle/water-rock reactions (weathering), as well as from pollutant sources such as mining, land clearance, agriculture, acid precipitation, domestic and industrial wastes. Compositional relations among dissolved species can reveal the origin of solutes and the process that generated the observed water compositions.

The hydrochemical evolution of groundwater can be understood by plotting the major cations and anions in the Piper tri-linear diagram and Durov diagram. These diagrams are of significance in bringing out chemical relationships among spring water in more definite terms. Presentation of chemical analysis in graphical form makes system. interpretation of complex spring/hand pump water simpler and quicker These diagrams are useful in screening and sorting large number of chemical data, major cations and anions such as Na^+, K^+, Ca^{2+}, Mg^{2+}, HCO_3^-, SO_4^{2-} and Cl^-, which makes interpretation easier. The Aqua-Chem 4.0 computer software was used for plotting these diagrams.

4.1.1 Piper diagram

The different spring/hand pump water samples have been classified according to their chemical composition using the Piper tri-linear diagram in this study. Two triangular fields, plotted separately, form the percentage meq/l values of the cations Ca^{2+} and Mg^{2+} (alkaline earths) and Na^+ (alkali), and the anions HCO_3^- (weak acid) and SO_4^- and Cl^-(strong acid). The relationship between different water bodies and processes influencing their chemical composition was investigated through the use of major ion concentrations versus Cl^- concentration because chlorine is the good indicator for mixing phenomena. Deviation from evaporation line indicates addition or removal of solutes as result of chemical reactions, such as mineral dissolution or precipitation.

The Piper tri-linear diagram is used to infer hydrogeochemical facies. The term "hydrogeochemical facies" is used to describeoccurrence modes of groundwater in an aquifer that differs in their chemical composition and identify the water composition in differentclasses. The facies are a function of lithology, solution kinetics, andflow patterns of the aquifer and recognizable parts of different characters belonging to any genetically related Hydrochemical facies are distinct zones that possess cation and anion concentration categories and this concept helps to understand and identify the water composition in different classes. To define composition class, Back and Hanshaw suggested subdivisions of the tri-linear diagram.

Classification diagram for anion and cation facies in the form of major-ion percentage is represented in figure 4.1. Water types are designed according to the domains in which they occur on the diagram segments.

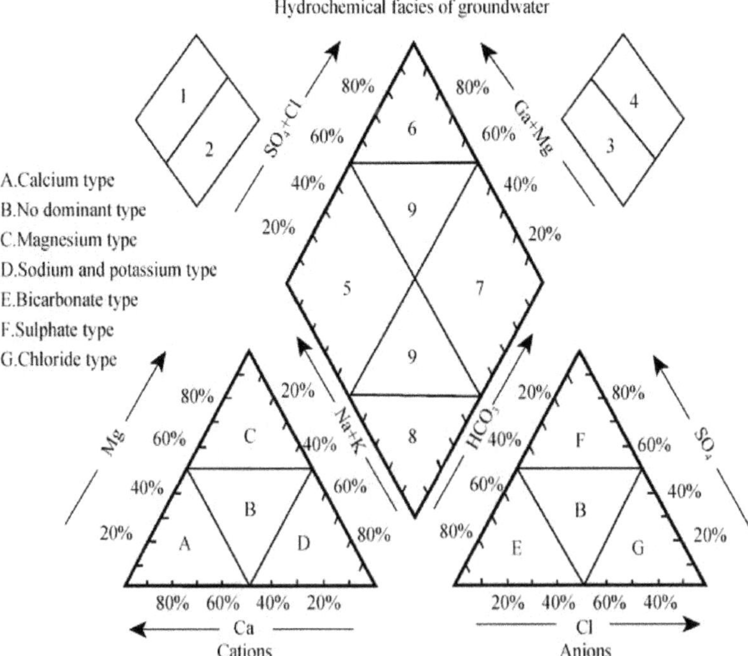

Fig. 4.1: Classification of Piper-trilinear diagram in the form of percentage

Based on the chemical characteristics, the spring/hand pump water have been categorized into three major groups - Calcium-Sodium-Chloride-bicarbonate (27% of the total samples), Calcium-Sodium-Bicarbonate-Chloride (21% of the total samples), and Sodium-Bicarbonate-Chloride (11% of the total samples). The overall water types are tabulated in table 4.1. The Piper trilinear diagrams, depicted in figure 4.2 to 4.4, stated that the majority of samples analyzed seasonally during 2007 and 2008 fall in the category of no dominance area. Figure 4.2 (a) (Pre-Monsoon 2007) demonstrates that the water samples occupy a wide area of the Piper diagram ranging from fresh to saline waters. 75% samples fall in the Ca^{2+}, HCO_3^- region and remaining 25% samples fall in no dominance area. In Pre-monsoon two major facies are dominant in the region, these are: (I) Ca-Na-Cl-HCO_3 (II) Ca-HCO_3-Cl.

The Figures 4.2 (b) (monsoon 2007) and 4.3 (a) (Post-monsoon 2007) shows that the majority of samples fall in the Ca^{++} and HCO_3^- region. The major water types found in Pre-monsoon 2008 are Ca-Na-Cl-HCO_3, Ca-Na-Cl and Na-Ca-HCO_3 (figure 4.3 (b)). However, monsoon 2008 figure 4.4 (a) cations-anions composition exhibits pattern similar to monsoon 2007 (figure 4.3 (a)). An examination of figure 4.4 (b) states that maximum samples have the mixed group effect of all the anions and cations, dominant facies in the area is Ca-Na-Cl-HCO_3. Piper-tri linear diagrams for the spring/hand pumps clearly supports that they fall in fresh water domain and these findings are substantiated by Durov diagrams.

Table 4.1: Spring/hand pump water type during pre monsoon, monsoon, and post monsoon 2007-2008

Sample ID Water type		A1	A2	A3	A4	A5	A6	A7
	Pre-Mon.	Ca-Na-Cl-HCO3	Ca-Na-Cl-HCO_3	Ca-Na-HCO_3-Cl	Ca-Na-HCO_3-Cl	Ca-Na-Cl-HCO_3	Ca-Na--Cl-HCO_3	Ca-Na-ClHCO_3
2007	Mon.	Ca-Na-Cl-HCO_3	Ca-Na-Cl-HCO_3	Ca-Na-Cl-HCO_3	Ca-Na-Cl-HCO_3	Ca-Na-Mg-Cl-HCO_3	Ca-Na-Cl-HCO_3	Ca-Na-Mg-Cl-HCO_3
	Post-Mon.	Ca-Na-Cl	Mg-Ca-Na-HCO_3	Ca-Mg-Na-Cl	Ca-Na-Mg-Cl-HCO_3	Ca-Na-Cl-HCO_3	Ca-Na-Cl-HCO_3	Ca-Na-Cl-HCO_3
	Pre-Mon.	Ca-Na-Cl-HCO_3	Ca-Na-Cl-HCO_3	Ca-Na-HCO_3-Cl	Ca-Na-HCO_3-Cl	Ca-Na-Cl-HCO3	Ca-Na-Cl-HCO_3	Ca-Na-Cl-HCO_3
2008	Mon.	Ca-Na-Cl-HCO_3	Ca-Na-Cl-HCO_3	Ca-Na-Cl-HCO_3	Ca-Na-Cl-HCO3	Ca-Na-Mg-Cl-HCO_3	Ca-Na-Cl-HCO_3	Ca-Mg-Na-Cl-HCO_3
	Post-Mon.	Ca-Na-Cl-HCO_3	Ca-Na-Mg-Cl-HCO_3	Ca-Na-Cl-HCO_3	Ca-Na-Mg-Cl-HCO_3	Ca-Na-Cl-HCO_3	Ca-Na-Cl-HCO_3	Ca-Na-Cl-HCO_3
Sample ID Water type		**A8**	**A9**	**A10**	**A11**	**A12**	**A13**	**A14**
	Pre-Mon.	Ca-Na-Cl-HCO3	Ca-Na-Cl-HCO_3	Ca-Na-HCO_3-Cl	Ca-Na-HCO_3-Cl	Ca-Na-Cl-HCO_3	Ca-Na-Cl-HCO_3	Ca-Na-Cl HCO_3
2007	Mon.	Ca-Na-Mg-Cl-HCO_3	Ca-Na-HCO_3-Cl	Ca-Mg-Cl-HCO_3	Ca-Na-Cl-HCO_3	Ca-Na-Cl-HCO_3	Ca-Na-Cl-HCO_3	Ca-Na-Mg-Cl-HCO_3
	Post Mon.	Ca-Na-Cl-HCO_3	Ca-Na-Mg-HCO_3-Cl	Ca-Mg-Na-HCO_3	Ca-Na-Cl-HCO_3	Ca-Na-Cl-HCO_3	Ca-Na-Cl	Ca-Na-Cl Ca-Na-Cl
	Pre-Mon.	Ca-Na-HCO_3-Cl	Ca-Na-HCO_3-Cl	Ca-HCO_3-Cl	Ca-Na-HCO_3-Cl	Ca-Na-HCO3-Cl	Ca-Na-Cl-HCO_3	Ca-Na-Cl
2008	Mon.	Ca-Na-Mg-Cl-HCO_3	Ca-Mg-HCO_3-Cl	Ca-Mg-Cl-HCO_3	Ca-Na-Mg-Cl-HCO3	Ca-Na-Cl-Cl-HCO_3	Ca-Na-Cl	Na-Ca-Cl-HCO_3
	Post Mon.	Ca-Na-Cl-HCO_3	Ca-Na-Cl-Cl-HCO_3	Ca-Mg-Na-HCO_3	Ca-Na-Mg-Cl-HCO_3	Ca-Na-Cl	Ca-Na-Ca-Na-Cl-HCO_3	Ca-Na-Cl-HCO_3

Table 3.7: Contd.

Sample ID	Water type	A15	A16	A17	M1	M2	M3	M4
	Pre-Mon.	Ca-Na-Cl-HCO3	Ca-Na-Cl	Ca-Cl-HCO_3	Ca-Na-HCO_3-Cl	Ca-HCO_3-Cl	Ca-HCO_3	Ca-HCO_3-Cl
2007	Mon.	Ca-Na-Cl	Ca-Na-Cl-HCO_3	Ca-Cl-HCO_3	Ca-HCO_3-Cl	Ca-HCO_3-Cl	Ca-Cl-HC	Ca-Cl- $Hcom_3$
	Post-Mon.	Ca-Na-Cl	Ca-Na-Mg-Cl-HCO_3	Ca-Na-Cl-HCO_3	Ca-Mg-HCO_3-Cl	Ca-HCO_3	Ca-Cl-HC	Ca-Mg-Cl-HCO_3
	Pre-Mon.	Ca-Na-Cl-HCO_3	Ca-Na-Cl-HCO_3	Ca-Na-Cl-HCO_3	Ca-HCO_3-Cl	Ca-HCO_3	Ca-Mg-c	Ca-HCO_3-Cl
2008	Mon.	Ca-Na-Cl	Ca-Na-Cl-HCO_3	Ca-Cl-HCO_3	Ca-HCO_3-Cl	Ca-Cl-HCO_3	Ca-Cl-HC	Ca-Cl-HCO_3
	Post-Mon.	Ca-Na-Cl	Ca-Na-Cl-HCO_{33}	Ca-Na-Cl-HCO_3	Ca-HCO_3-Cl	Ca-HCO_3-Cl	Ca-Cl-HC	Ca-Mg-Cl-HCO_3

Sample ID	Water type	M5	M6	S1	S2	S3	S4	S5
	Pre-Mon.	Ca-HCO_3	Ca-Na-HCO3-SO_4-Cl	Ca-Na-HCO_3-Cl	Ca-Na-HCO_3-Cl	Ca-Na-HCO_3	Ca-Na-HCO_3-Cl	Ca-HCO_3-Cl
2007	Mon.	Ca-HCO_3-ClSO_4	Ca-HCO_3-Cl	Na-Ca-HCO_3	Ca-HCO_3-Cl	Ca-Na-HCO_3-Cl	Ca-Cl-HCO_3	Ca-HCO_3-Cl
	Post-Mon.	Ca-Mg-HCO_3	Ca-Mg-Na-Cl-HCO_3-SO_4	Na-Ca-HCO_3	Ca-Na-Mg-HCO_3	Ca-Na-HCO_3	Ca-Na-HCO_3-Cl	Ca-HCO_3-Cl
	Pre-Mon.	Ca-HCO_3-Cl	Ca-Na-HCO_3-Cl-SO_4	Ca-Na-HCO_3--Cl	Ca-Na-HCO_3-Cl	Ca-Na-HCO_3-Cl	Ca-	Ca-Cl-HCO_3
2008	Mon.	Ca-HCO_3-Cl	Ca-Na-HCO_3-SO_4	Na-Ca-Mg-HCO_3	Ca-HCO_3-Cl	Ca-Na-HCO_3-Cl	Ca-Cl-HCO_3	Ca-HCO_3-Cl
	Post-Mon.	Ca-Mg-HCO_3	Ca-Mg-Na-HCO_{33-}Cl-SO_4	Na-Ca-HCO_3	Ca-Na-Mg-HCO_3-Cl	Ca-Na-HCO_3-Cl	Ca-Cl-HC	Ca-HCO_3-Cl

Table 3.7: Contd.

Sample ID	Water type	S6	S7	S8	P1	P2	P3	P4
	Pre-Mon.	Ca-Na-HCO_3-Cl	Ca-HCO3-SO_4-Cl	Ca-HCO_3-Cl	Ca-Na-HCO_3-Cl	Ca-Mg-Na-HCO_3	Ca-Na-HCO_3	Na-Mg-Ca-HCO_3-Cl
2007	Mon.	Ca-HCO_3-ClSO_4	Ca-HCO_3-Cl	Na-Ca-HCO_3	Ca-HCO_3-Cl	Ca-Na-HCO_3-Cl	Ca-Cl-HCO_3	Ca-Na-HCO_3-Cl
	Post-Mon.	Ca-Mg-HCO_3	Ca-Mg-Na-Cl-HCO_3-SO_4	Na-Ca-HCO_3	Ca-Na-Mg-HCO_3	Ca-Na-HCO_3	Ca-Na-HCO_3-Cl	Ca-Na-Mg-HCO_3
	Pre-Mon.	Ca-HCO_3-Cl	Ca-Na-HCO_3-Cl-SO_4	Ca-Na-HCO_3--Cl	Ca-Na-HCO_3-Cl	Ca-Na-HCO_3-Cl	Ca-	Ca-Na-HCO_3-Cl
2008	Mon.	Ca-Na-Cl-HCO_3	Ca-HCO_3-SO_4-Cl	Na-HCO_3-Cl	Ca-Na-HCO_3-Cl-SO_4	Ca-Cl-HCO_3	Ca-Na-HCO_3	Ca-Na-HCO_3-Cl
	Post-Mon.	Na-Ca-Mg-HCO_3-Cl	Ca-HCO_{33}-SO_4-Cl	Ca-Mg-HCO_3	Ca-Na-HCO_3-Cl	Ca-Mg-HCO_3	Ca-Mg-HCO_3	Ca-Na-Mg-HCO_3-Cl

Sample ID	Water type	P5	P6	P7	J1	J2	J3	J4
	Pre-Mon.	Ca-Na-HCO_3-Cl	Ca-Na-HCO_3	Na-Ca-HCO_3-SO_4	Ca-Na-HCO_3-Cl	Ca-HCO_3-Cl	Ca-Na-HCO_3	Ca-Na-HCO_3-Cl
2007	Mon.	Ca-Cl-HCO_3	Ca-Na-Cl-HCO_3	Ca-Na-HCO_3-SO_4	Ca-Cl-HCO_3	Ca-HCO_3-Cl	Ca-Na-HCO_3	Ca-HCO_3-Cl
	Post-Mon.	Ca-Mg-HCO_3	Ca-Mg-Na-Cl-HCO_3-SO_4	Na-Ca-HCO_3	Ca-Na-Mg-HCO_3	Ca-Na-HCO_3	Ca-Na-HCO_3-Cl	Mg-Ca-Na-HCO_3
	Pre-Mon.	Ca-Na-HCO_3-Cl	Ca-Na-HCO_3-Cl	Na-Ca-HCO_3-Cl	Ca-Na-HCO_3-Cl	Ca-HCO_3-Cl	Ca-HCO_3	Ca-Na-HCO_3-Cl
2008	Mon.	Ca-Cl-HCO_3	Ca-Na-HCO_3-Cl	Ca-Na-HCO_3-SO_4	Ca-Cl-HCO_3	Ca-HCO_3-Cl	Ca-Na-HCO_3	Ca-HCO_3-Cl
	Post-Mon.	Ca-Na-Cl-HCO_3	Ca-Na-Mg-HCO_{33}-Cl	Ca-Na-HCO_3-SO_4	Ca-Na-HCO_3-Cl	Ca-Na-HCO_3-SO_4	Ca-Mg-HCO_3	Ca-Na-HCO_3

Table 3.7: Contd.

Sample ID	Water type	J5	J6	HP1	HP2	HP3	HP4	HP5
	Pre-Mon.	Ca-HCO_3	Ca-Na-HCO_3-Cl	Ca-HCO_3-Cl	Ca-Mg-HCO_3-Cl	Ca-Mg-HCO_3-Cl	Ca-Mg-HCO_3-Cl	Ca-HCO_3-Cl-SO_4
2007	Mon.	Ca-Na-HCO_3-Cl	Ca-Na-HCO_3	CaHCO_3-Cl	Ca-MgHCO_3-C_l	Ca-MgHCO_3	Ca-Na-HCO_3-Cl	Ca-HCO_3-SO_4-Cl
	Post-Mon.	Ca-Mg-HCO_3	Ca-Na-Mg-HCO_3	Ca-HCO_3	Ca-Na-HCO_3	Ca-HCO_3	Ca-Na-HCO_3	Na-Ca-SO_4-Cl-HCO_3
	Pre-Mon.	Ca-HCO_3	Ca-Na-HCO_3-Cl	Ca-HCO_3	Ca-Na-HCO_3-Cl	Ca-HCO_3	Ca-HCO_3	Ca-Na-HCO_3
2008	Mon.	Ca-Na-HCO_3-Cl	Ca-Na-HCO_3-Cl-SO_4	Ca-Cl-HCO_3	Ca-Mg-HCO_3-Cl	Ca-Mg-HCO_3	Ca-Na-HCO_3-Cl	Ca-HCO_3-SO_4
	Post-Mon.	Ca-Mg-Na-HCO_3	Ca-Mg-Na-HCO_3	Ca-HCO_3-Cl	Ca-Na-HCO_3	Ca-HCO_3	Ca-Na-HCO_3	Na-HCO_3

Sample ID	Water type	HP6	HP7	HP8	HP9	HP10
	Pre-Mon.	Ca-Na-HCO_3	Ca-Mg-HCO_3-Cl	Ca-Na-HCO_3-Cl	Ca-Mg-HCO_3-Cl	Ca-Mg-HCO_3-Cl
2007	Mon.	Ca-Na-HCO_3	HCO_3-Cl	Ca-Na-HCO_3-Cl	Ca-Mg-HCO_3-Cl	Ca-MgHCO_3
	Post-Mon.	Ca-Na-HCO_3	Ca-Na-HCO_3	Ca-Na-HCO_3-Cl	Mg-Ca-HCO_3-Cl	Ca-Mg-HCO_3-Cl
	Pre-Mon.	Ca-Na-HCO_3	Ca-HCO_3	Ca-Na-HCO_3	Ca-HCO_3-Cl	Ca-HCO_3-Cl
2008	Mon.	Ca-Na-HCO_3-Cl	Ca-Mg-HCO_3-Cl	Ca-Na-HCO_3-Cl	Ca-Mg-Cl-HCO_3	Ca-Mg-HCO_3-Cl
	Post-Mon.	Ca-Na-HCO_3	Ca-Mg-HCO_3	Ca-Na-HCO_3-Cl	Mg-Ca-HCO_3-Cl	Ca-Mg-HCO_3-Cl

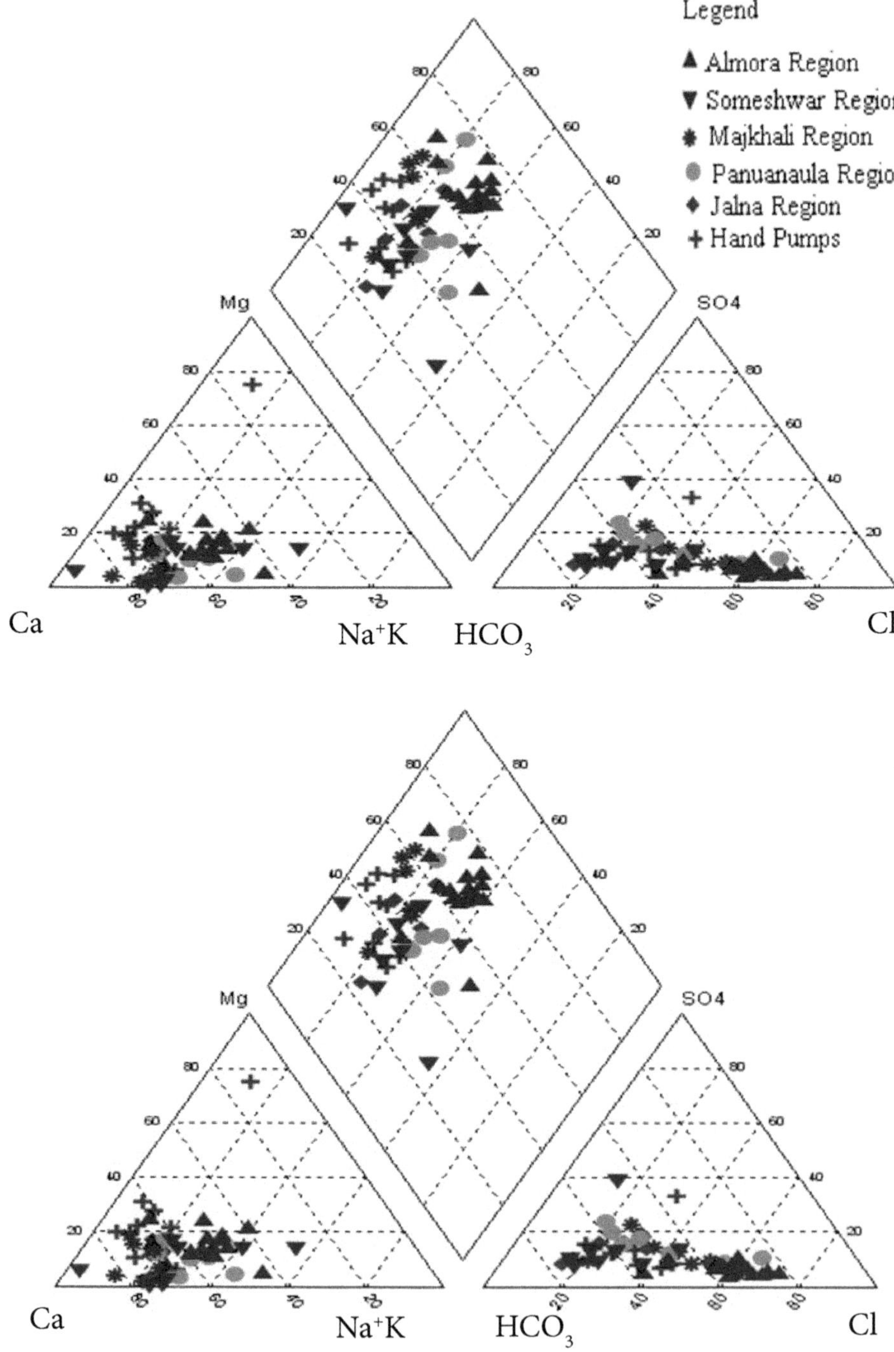

Fig. 4.2: Piper-tri linear diagram of spring/hand pump water during (a) pre monsoon 2007, and (b) monsoon 2007

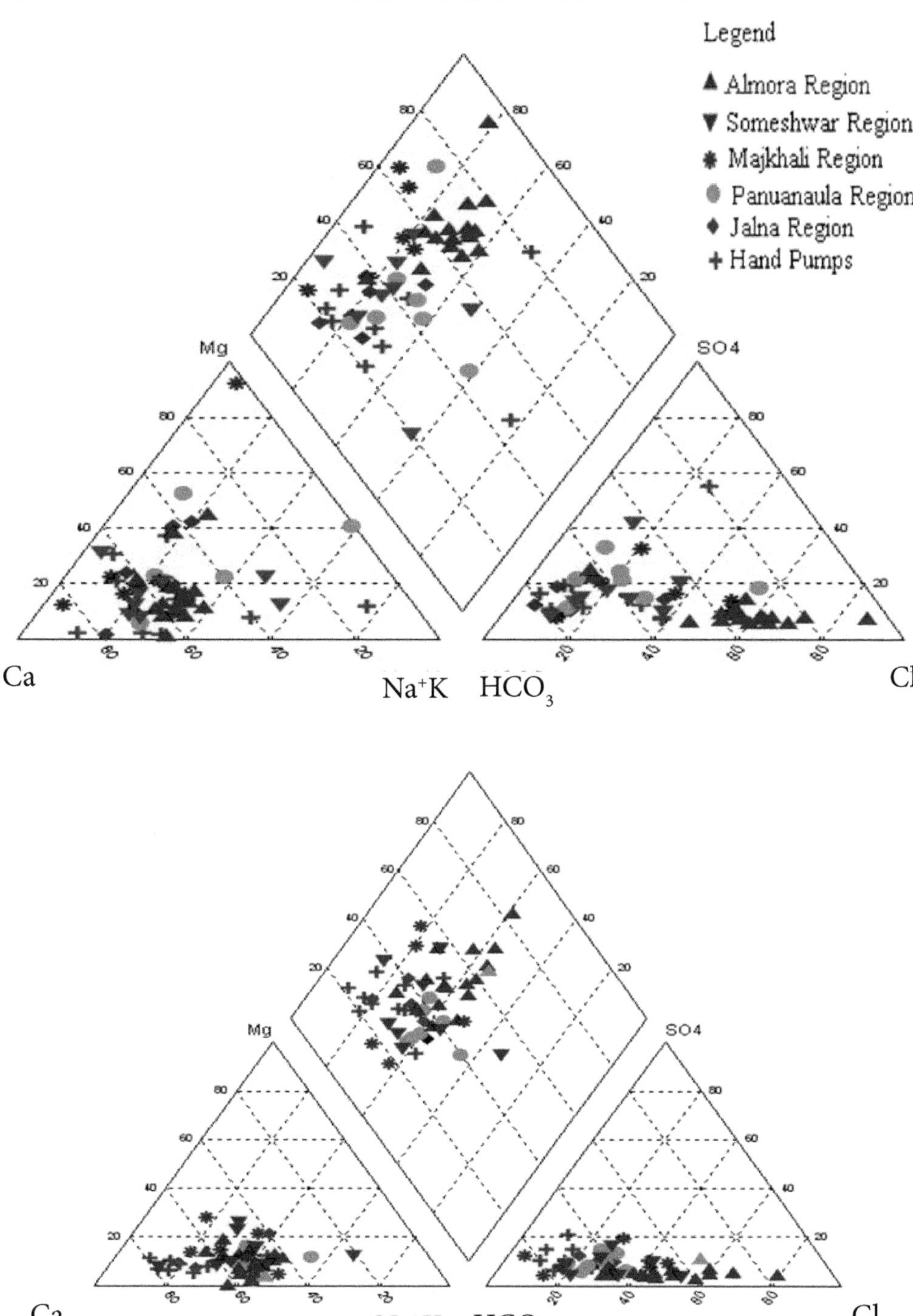

Fig. 4.3: Piper-tri linear diagram of spring/hand pump water during (a) post monsoon 2007, and (b) pre monsoon 2008

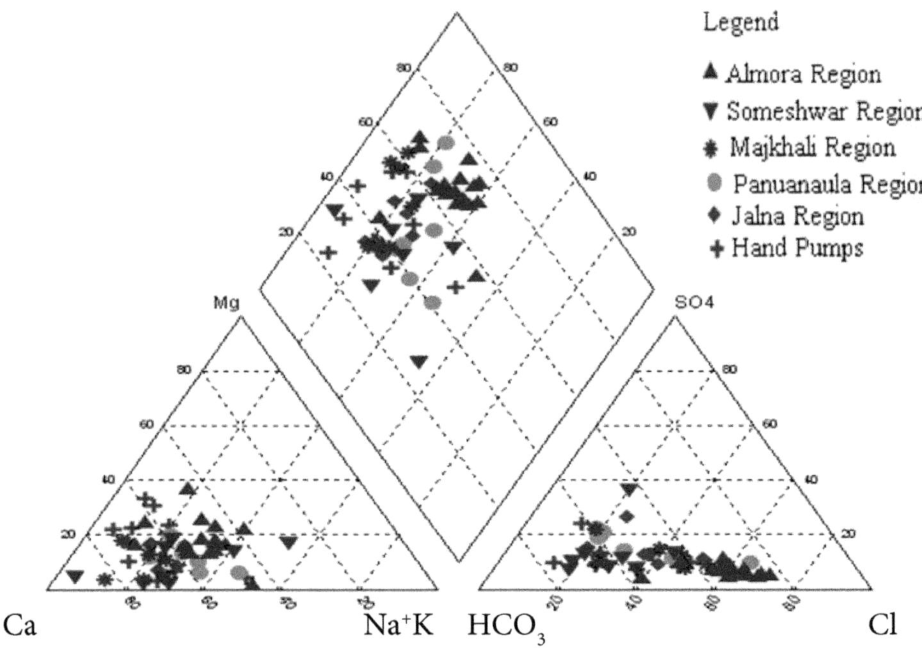

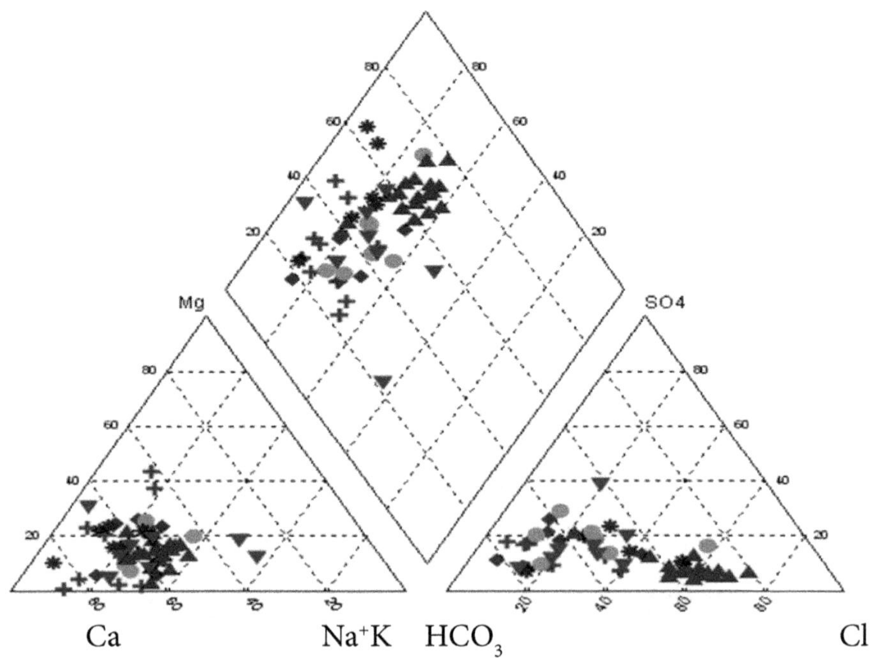

Fig. 4.4: Piper-tri linear diagram of spring/hand pump water during (a) monsoon 2008, and (b) post monsoon 2008

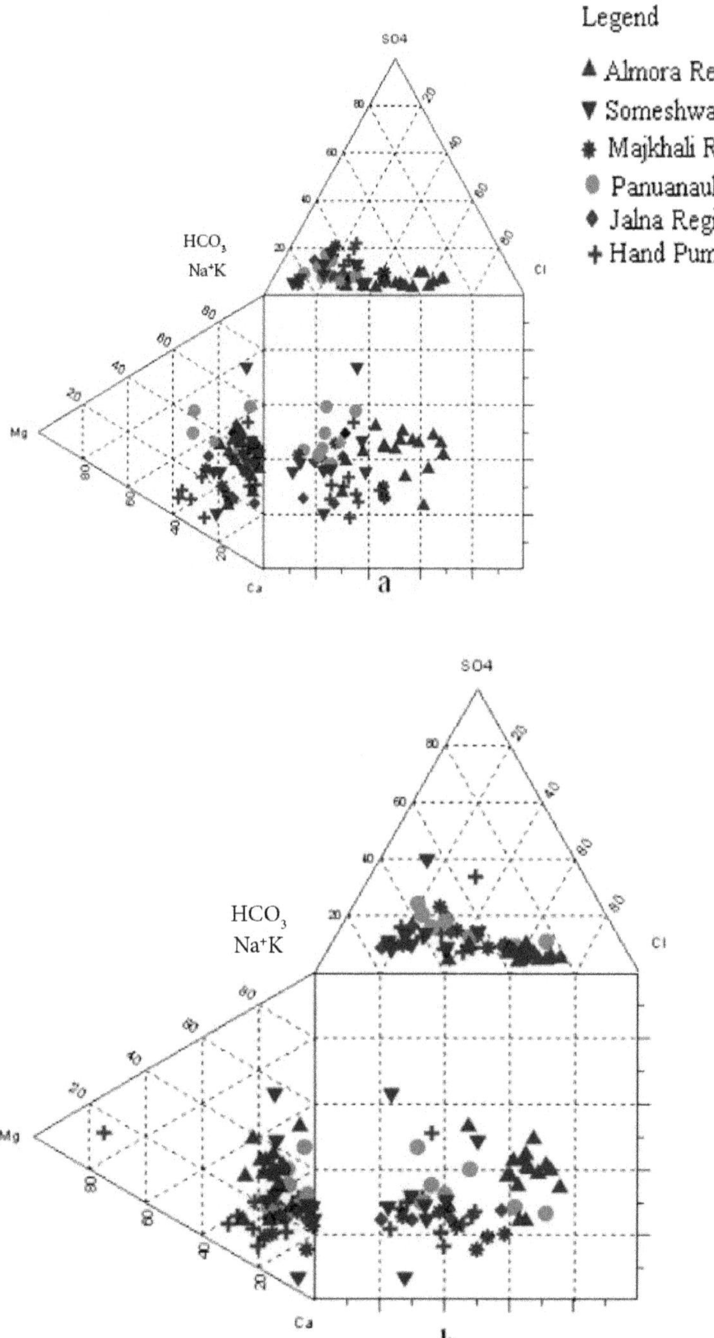

Fig. 4.5: Durov diagram of spring/hand pump water during (a) premonsoon 2007, and (b) monsoon 2007

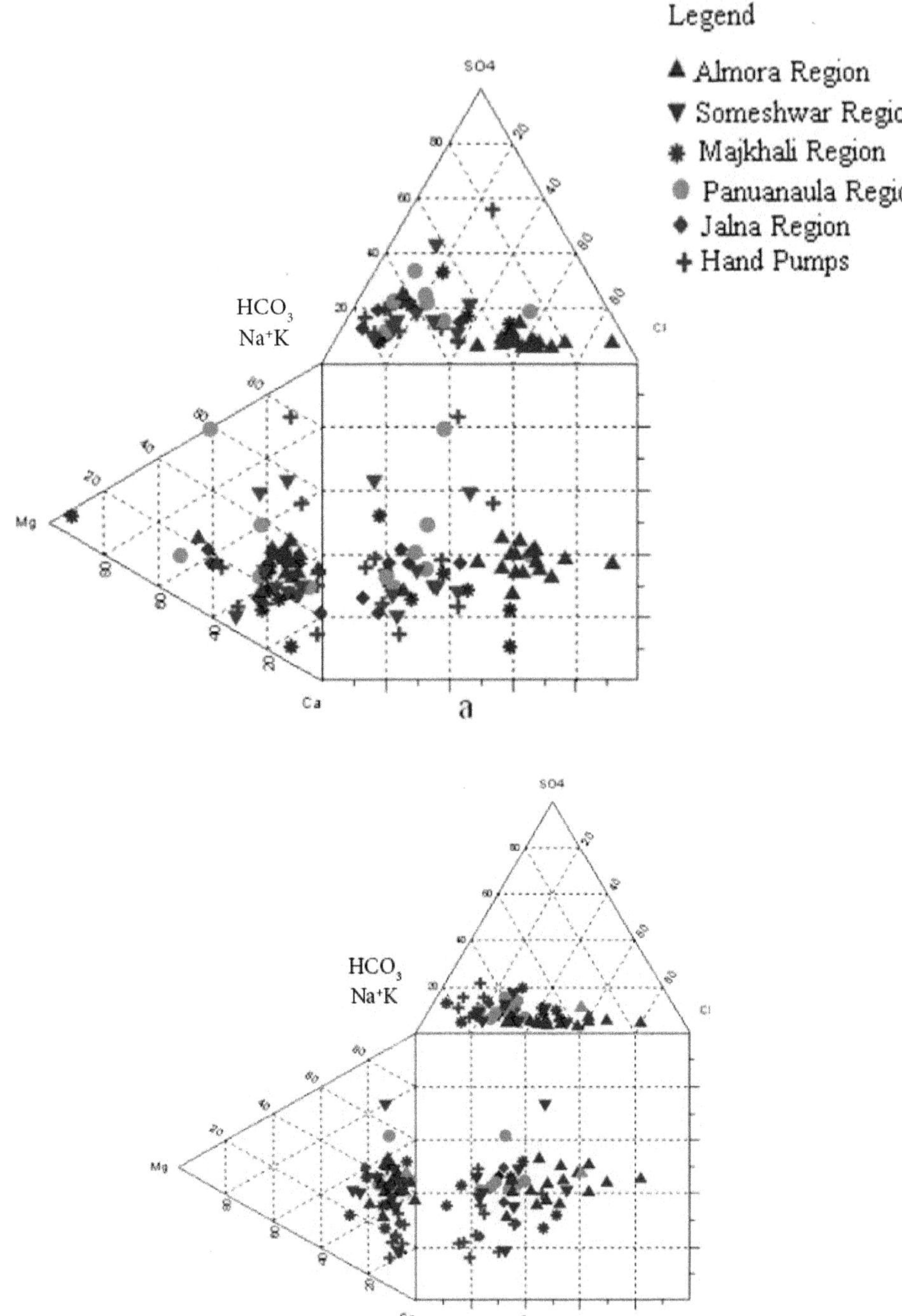

Fig. 4.6: Durov diagram of spring/hand pump water during (a) postmonsoon 2007, and (b) pre monsoon 2008

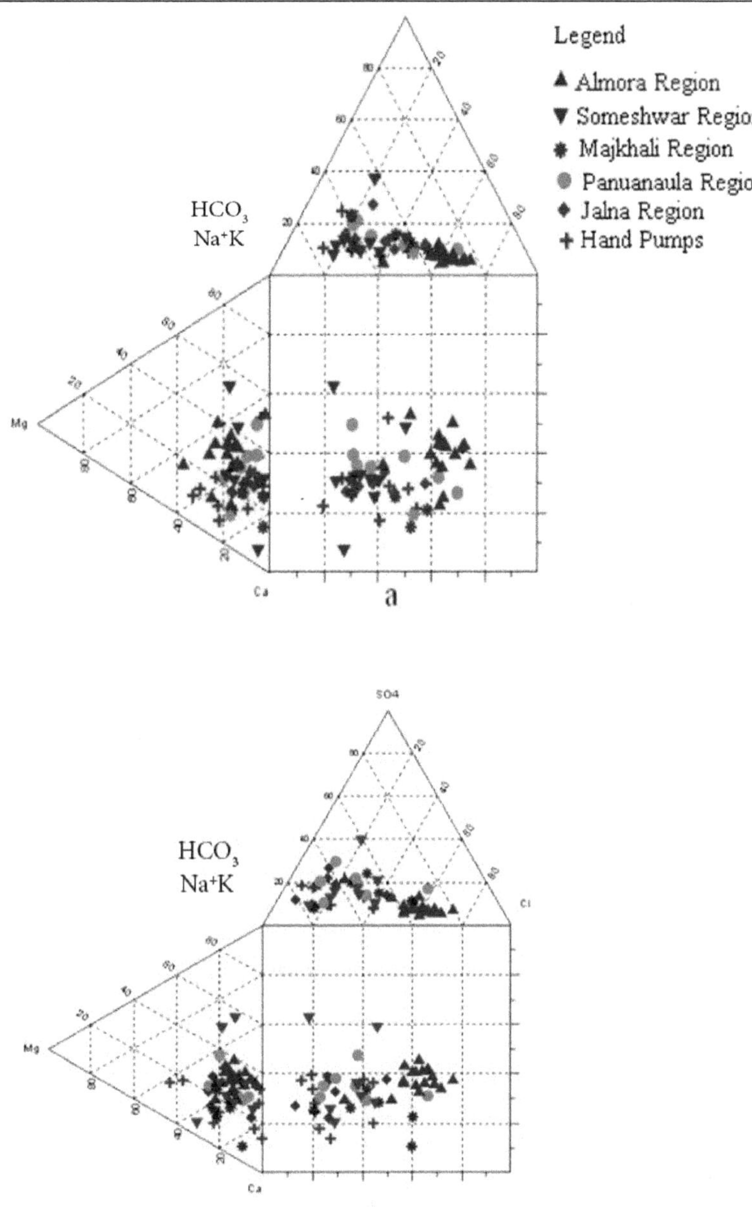

Fig. 4.7: Durov diagram of spring/hand pump water during (a) monsoon 2008, and (b) post monsoon 2008

4.1.2 Durov diagram

Durov diagram is based on the percentage of the major ions' in meq/L. Both the positive and the negative ion percentages total 100 %. The values of the cations and the anions are plotted in the appropriate triangular and projected into the square of the main field. The advantage of this diagram is that it displays some possible geochemical processes that could affect the water genesis. Durov diagrams are presented in figure 4.5 to 4.7 during pre monsoon, monsoon, and post monsoon during 2007 and 2008.

4.2 Hydro-chemical characterization of springs/hand pumps

The physico-chemical parameters of springs/hand pump of different features of Almora district have been studied during 2007 and 2008. Standards values as per BIS specification and WHO guidelines for water quality characteristics/parameters are tabulated in table 4.2. Springs/hand pumps often consists eight major chemical ions, major cations - calcium, magnesium, sodium, and potassium and major anions - bicarbonate, chloride, nitrate and sulfate.

4.2.1 Temperature

Temperature readings are used in the calculation of various forms of alkalinity, salinity and calcium, in studies of saturation and stability with respect to calcium carbonate, and in general laboratory operations. Warmer temperatures decrease the solubility of calcium carbonate. Low temperatures decrease the solubility of barium sulfate, strontium sulfate, and silica. Similar observations were made by Verlecar and coworker40. The temperature variation is found in between 1.2 to 20, 0.6 to 12.1, and -6.4 to 22.8 ^{0}C during pre-monsoon, monsoon, post-monsoon 2007 respectively. During pre-monsoon, monsoon, post-monsoon 2008, the temperature variation recorded is 12 to 21, 0.8 to 13.1, and 7.1 to 20.2^{0}C respectively.

4.2.2 pH

pH is a method of expressing hydrogen ion concentration. pH determines whether the water is acidic or alkaline. To the water chemist, pH is important in defining the alkalinity equilibrium levels of carbon dioxide, bicarbonate, carbonate and hydroxide ions. The pH of most natural waters ranges from 6.5 to 8.5. Deviation from the neutral pH 7.0 is largely the result of interaction between acids and bases.

Industrial and community wastes, acid rain, bedrock type, and the biological processes of photosynthesis and respiration influence the pH level. As rain water falls, it dissolves carbon dioxide from the atmosphere, thus forming a weak carbonic acid and lowering the pH of the precipitation. The pH of water is extremely important to aquatic life. Most aquatic species tolerate a limited pH range and most fish require a pH above 5.5 for growth and reproduction.

Low pH levels can have a harmful impact on the health of aquatic communities. The lower pH value tends to make water corrosive and higher pH results in the taste complaint and can have negative impact on skin and eyes. Very acidic water or acid rain can allow toxic substances, such as ammonia and heavy metals like lead to leach from the rocks and possibly be taken up by aquatic plants and animals (bioaccumulation).

Table 4.2: Spring/hand pump water quality of the studied area withWHO and BIS standards

Parameters	World Health Organization		Bureau of Indian Standards	
	Permissible	Excessive	Permissible	Excessive
pH	6.5-8.5	6.5-9.2	7.0-8.5	7.0-9.2
EC	-	-	0.1-5.0	0.1-6.0
DO	4.6-6.0	-	-	-
TH	300	600	100	500
Bicarbonate	300	600	300	600
Calcium	75	200	75	200
Magnesium	30	100	50	150
Alkalinity	200	600	200	600
Nitrate	45	-	45	-
Chloride	200	600	250	1000
TDS	500	1500	500	2000
Sulfate	200	-	200	400
Iron	0.3	1.0	0.3	1.0
Chromium	0.05	-	0.05	-
Copper	0.05	-	0.05	1.5
Zinc	5	-	5	15

The pH values that exceed 9.0 cause excessive algal growth, a sign of nutrient enrichment. Very low pH readings are generally observed near the point sources of pollution. pH of the springs/hand pumps lies in the range of $5.38<pH<8.03$ seasonally during 2007 and 2008 but approximately 85% of the total samples were slightly acidic in nature.

4.2.3 ORP (oxidation reduction potential)

pH directly relates with the Eh and it increases with decreasing Eh. In the presentation, similar trends have been observed *i.e.* springs with low Eh exhibit high pH and spring with low pH present high Eh value (Figure 4.8). Redox (Eh) potential is found within the range of -24 to129 and -24 to 120 mv during pre-monsoon, monsoon, post-monsoon 2007 and 2008 respectively.

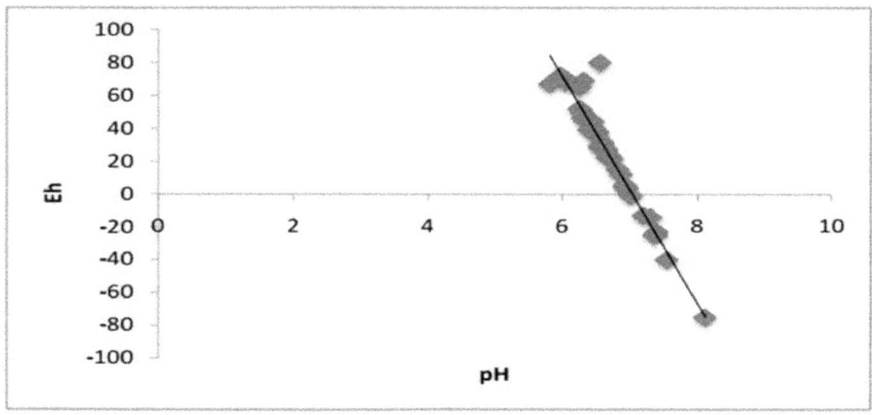

Fig. 4.8: Variation of pH with Eh (Pre monsoon 2007)

4.2.4 Electrical Conductivity (EC)

Electrical Conductivity (EC) is a measure of the ability of water to conduct electrical current. Conductivity measures the amount of ions in a solution. The more the ions in the solution the higher is the conductivity. This means that the eutrophic water has higher conductivity than oligotrophic water. Thus, conductivity in water is affect by the presence of inorganic dissolved solids such as chloride, nitrate, sulfate and phosphate anions or sodium, magnesium, calcium, iron and aluminium cations.

The EC of water is an indirect measure of the total dissolved solids (TDS) content of water, and there is usually an approximately linear relationship between TDS and conductivity.

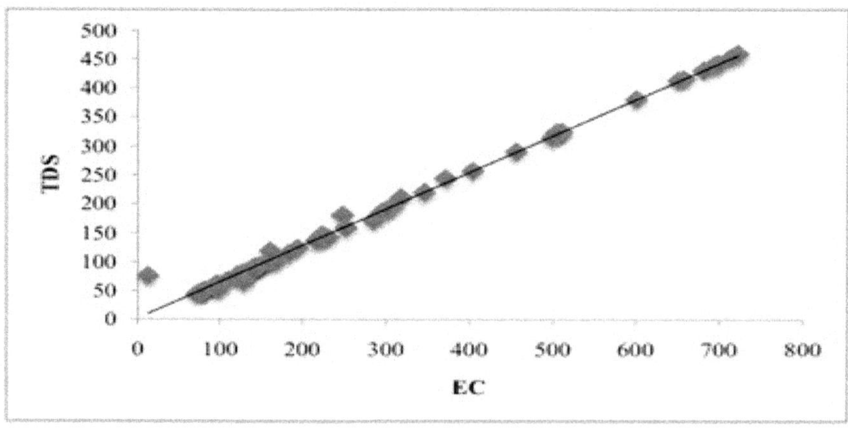

Fig. 4.9: Variation of EC with TDS (Pre monsoon 2007)

The conductivity is recorded higher in monsoon season than in post monsoon season and in pre-monsoon during two consecutive years 2007 and 2008, may be due to increased dissolution of salts in rain water. EC values for the springs are found between 66 to 749 and 63 to 762µS/cm during 2007 and 2008 respectively. TDS of the representative samples ranges from 0.07 to 464 mg/l. The linear relation between EC and TDS is presented in the figure 4.9.

4.2.5 Dissolved Oxygen (DO)

Dissolved oxygen (DO) values show the ability of the groundwater to purify itself through biochemical process. Moving water, because of its churning, dissolves more oxygen than still water. Respiration by aquatic animals, decomposition of organic matter, and various chemical reactions consume DO. Thus, DO is an important water quality parameter to assess the waste assimilative capacity of the waters. Oxygen is less soluble in warmer waters and at higher altitudes where atmospheric pressure is lower. To determine the percent oxygen saturation of water, temperature and DO data are needed. As the temperature of water rises, metabolic demand for oxygen increases. Since oxygen is less soluble in warmer waters, increased temperatures coupled with greater demand can create critically low oxygen levels. If more oxygen is consumed than is produced, DO level decline and some sensitive animals may move away, weaken or die. DO levels fluctuate seasonally, daily, and with water temperature. Aquatic animals are the most vulnerable to lower DO levels in the early morning on hot summer days when flows are low, water temperatures are high, and aquatic plants have not been producing oxygen since sunset. During the study period the average DO observed in the range from 3.26 to 13.8 mg/l. The increase of temperature, decrease in DO is observed in the present study and variation of DO with water temperature (Pre monsoon 2007) shown in figure 4.10.

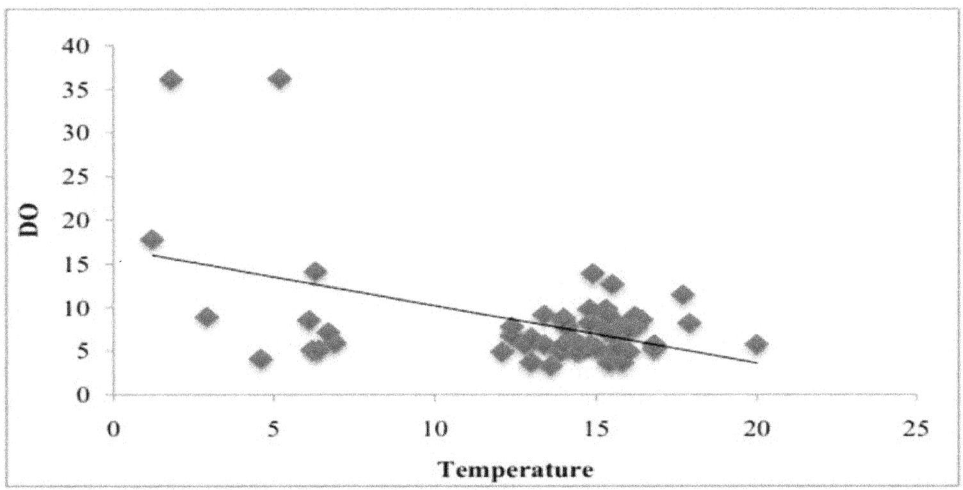

Fig. 4.10: Variation of DO with Water temperature (Pre monsoon 2007)

4.2.6 Hardness

The term "hardness" is one of the oldest terms used to describe characteristics of water. In fact, Hippocrate (450-34 BC) used the "Hard" and "Soft" terms in a discourse on water quality. He stated "Consider the waters which the inhabitants use, whether they be marshy and soft, or hard and running from elevated and rocky situation...". Hardness of the water is caused primarily by calcium and magnesium, although iron and magnesium also contribute to the actual hardness. Calcium and magnesium enter the water via the action of carbonic acid. As water and carbon dioxide react, carbonic acid is produced and dissolves calcium and magnesium from carbonate rocks (*e.g.* limestone, dolomite). Hardness may be divided into two types: carbonate and non carbonate. Carbonate hardness is that portion of calcium and magnesium that can combine with bicarbonate to form calcium and magnesium carbonate. A general accepted classification for hardness as $CaCO_3$ is, as follows:

S.No.	Hardness Range	Description
1.	0 to 60	Soft
2.	61 to 120	Moderately Hard
3.	121 to 180	Hard
4.	181 and above	Very Hard

Hard water increases the amount of soap or detergent needed to lather and leaves scaly deposits on containers and plumbing. Hardness levels above 500mg/l are not desirable for domestic use.

The hardness is found to increase during monsoon due to increased dissolution of CO_2 and minerals like Ca and Mg in rain water. Hardness varies from 18 to 208, 23 to 131, and 32 to 260 mg/l in 2007 and from 16 to 230, 26 to 244, and 36 to 365 in 2008 during pre-monsoon, monsoon, post-monsoon respectively.

4.2.7 Alkalinity

Alkalinity is a measure of the capacity of water to neutralize acids. It is primarily determined by the presence of carbonate, carbonates and hydroxides in water. These alkaline compounds in the water remove H^+ ions and lower the acidity of the water (increased pH). They usually do this by combining with the H^+ ions to make new compounds. At the pH of most irrigation water, alkalinity is primarily a measure of bicarbonate in the water. The alkalinity is also a measure of its buffering capacity. The higher the value, the more acid can be neutralized. Alkalinity of natural waters is primarily the result of bicarbonates but is expressed in terms of calcium carbonate. Alkalinity varies from 22.39 to 180, 20 to 13, and 3 to 136 mg/l in 2007 and 20 to 188, 20 to 138, and 16 to 138 mg/l in 2008 during pre-monsoon, monsoon, post-monsoon respectively.

4.2.8 Calcium

The presence of calcium in water supplies results from passage through or over deposits of limestone, dolomite and gypsum. Small concentrations of calcium combat corrosion of metal pipes by lying down a protective coating. Appreciable calcium salts, on the other hand, precipitate on treating to form harmful scale in boilers, pipes and cooking utensils. Calcium varies from 6.06 to 75.24, 10 to 79.48, and 0.43 to 78.22 mg/l in 2007 and 5.78 to 75.72, 9.25 to 79.08, and 10.42 to 80.21 mg/l in 2008 during pre-monsoon, monsoon, post-monsoon respectively.

4.2.9 Magnesium

Magnesium ranks eighth among the elements in order of abundance and is a common constituent of natural water. Mg of the springs/hand pumps lies in the range of $0.06<Mg<54.56$ and $0.3<Mg<29.52$ seasonally during 2007 and 2008 respectively.

4.2.10 Chloride

Chloride ion is one of the major inorganic anions in water. In potable water, the salty taste produced by chloride concentration is variable and dependent on the chemical composition of water. High chloride content may harm metallic pipes and structure, as well as growing plants. The main operational issue for chloride is its ability to the corrosiveness of water, particularly in low alkalinity water. Chloride concentration ranges between 6.73 from 113.46 mg/l in the springs/hand pumps of the study area.

4.2.11 Nitrate

Naturally high nitrate concentrations may occur in groundwater in semi arid or arid areas where there is widespread termite activity, or where natural vegetation is dominated by leguminous species such as acacias. Nitrate and nitrites are the most abundant forms of dissolved nitrogen in groundwater due to agricultural and domestic activities. Elevated levels are primarily associated with human contamination from fertilizers and sewage. High concentration of nitrates can stimulate the growth of aquatic plants and may be a health hazard to juvenile mammals. In the digestive system, nitrates are reduced to nitrites. As nitrates enter the bloodstream, hemoglobin is oxidized to methemoglobin rendering it incapable of transporting oxygen. Cyanosis due to methemoglobinemia (blue baby syndrome) may result from drinking water with high nitrate concentrations. The nitrate levels are increased during monsoon due to heavy rains, increased filling and chemical application of fertilizers to grow crops.

Nitrate concentration of Almora town is found slightly higher than its basic limit ranging from 1.06 to 40.79 mg/l. The concentration of nitrate of other region is found within the limit of 0.03-29.21 mg/l.

4.2.12 Cation-anion Relationship

Calcium and sodium were the major cations and individually accounted for 54-69% and 20-29% of the total cations respectively. The Ca^{2+}: Na^{+} equivalent ratio changes from 1.83 to 2.10. Bicarbonates accounts for 58-64% of total anions and varied from 29.89 to 185.24 mg/l. sulfate concentration is found between 5.99 and 25.94 mg/l. Potassium and Magnesium cations account for 3-7% and 6-8% of the total cations respectively. The Mg^{2+}: Ca^{2+} ratio vary from 0.09 to 0.16 and K^{+}: Na^{+} exhibit values between 0.15 to 0.24 and Cl^{-}: SO_4^{2-}equivalent recorded from 1.78 to 2.81.

The concentration of major ions usually followed the trend $HCO_3^{-}>Cl^{-}>Ca^{2+}>Na^{+}>SO_4^{2-}>NO_3^{-}>Mg^{2+}>K^{+}$.

4.3 Mechanisms controlling spring water chemistry

To know the groundwater chemistry and relationship of the chemical components of water from their respective aquifers (chemistry of the rock types, chemistry of precipitated water and rate of evaporation), Gibbs has developed a pictorial presentation in which ratios of dominant anions and cations are plotted against the values of total dissolved solids (TDS).

4.3.1 Gibbs diagrams

Gibbs diagrams, representing the ratio for cations [Na/(Na+Ca)] and ratio for anions [Cl/(Cl+HCO_3)] as a function of TDS are widely employed to assess the functional sources of dissolved chemical constituents for the ground water.

The dissolved ions in spring water are derived from various sources and compositional relations among them, can reveal the origin of solutes and the processes that generates the observed water compositions. The Na^+:Cl^- ratio has frequently been used to identify saline water intrusion. Based on the Na^+/Cl^-molar ratio, Meybeck interpreted the source of Na^+ in water as silicate weathering if the value is greater than 1.0. While comparing the aforesaid chemical quotient, it is found that all the values for the springs are less than 1.0 (except for some samples in Someshwar region) this simply states that the source of Na^+ is not due to the silicate weathering in the study area.

The plot of (Na^+ + K^+) vs. Cl^- (Fig.4.11A) shows that most of the values fall on and below the 1:1 line and suggests that the alkali is not balanced by the chloride ions. The interrelationship between HCO_3^- vs. Na^+ and HCO_3^- vs. Mg^{2+} (Fig.4.11 B and 4.11 C) shows that most of the data deviated from the expected 1:1 relation and this indicates for the multiple sources of Mg^{2+} and Na^+. The deviation from the 1:1 relation of Ca^{2+} vs. SO_4^{2-} indicates that dissolution of gypsum is not a major source of Ca^{2+}. The poor correlation between Mg^{2+} and SO_4^{2-} as well as between Na^+ and SO_4^{2-} suggests that dissolution of Na^+ and Mg^{2+} minerals is not the major source of sulfate in the spring water.

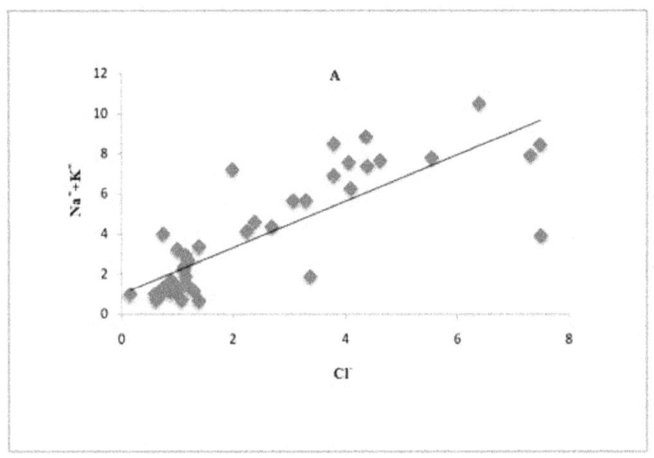

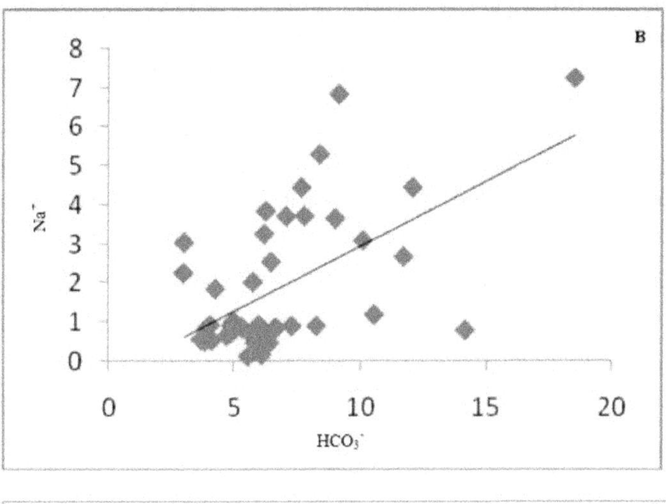

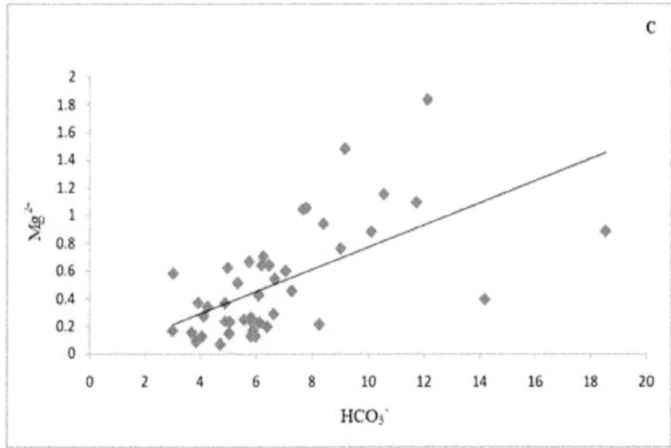

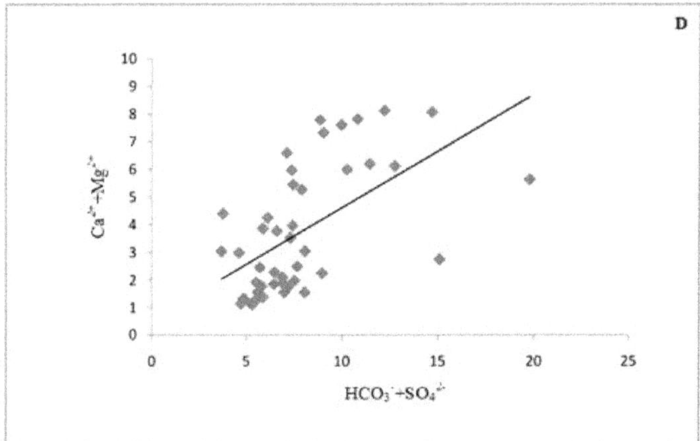

Fig. 4.11: Plot showing the interrelationship between (A) Na^++K^+vs. Cl^-, (B) HCO_3^- vs. Na+, (C) HCO_3^- vs. Mg^{2+} and (D) (Ca^{2+} + Mg^{2+}) vs. (HCO_3^- + SO_4^{2-}).

The deviation from the 1:1 line plot of (Ca^{2+} + Mg^{2+}) vs. (HCO_3^- + SO_4^{2-}) (Fig. 4.11 D) reveals that dissolution reactions of calcite, dolomite and gypsum are not dominant in the system. Ion exchange tends to shift the points to the right of the line due to an excess of HCO_3^- + SO_4^{2-}. If there was a large excess of Ca^{2+} + Mg^{2+} due to the reverse ion exchange process over HCO_3^- + SO_4^{2-}, then the points would have plotted to the left side of the line. This explanation was also supported by Gibbs diagram in the present study in which almost all samples lies within the rock dominance area.

The chemical data of groundwater samples are plotted in Gibbs diagram as presented in figures from 4.12 to 4.18.

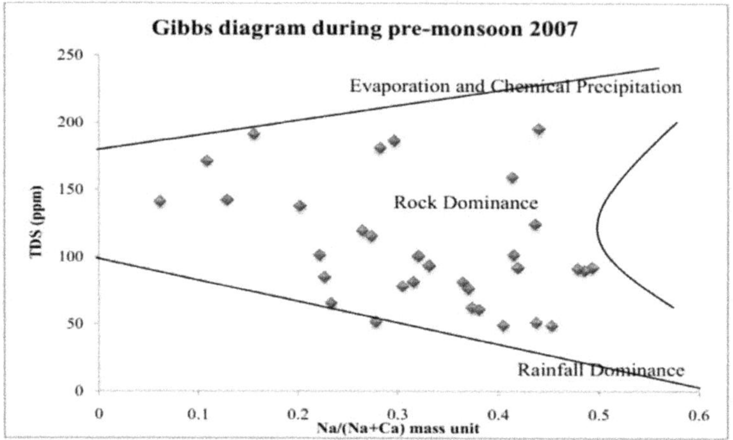

Fig. 4.12: Gibbs diagrams for the cations and anions of groundwater

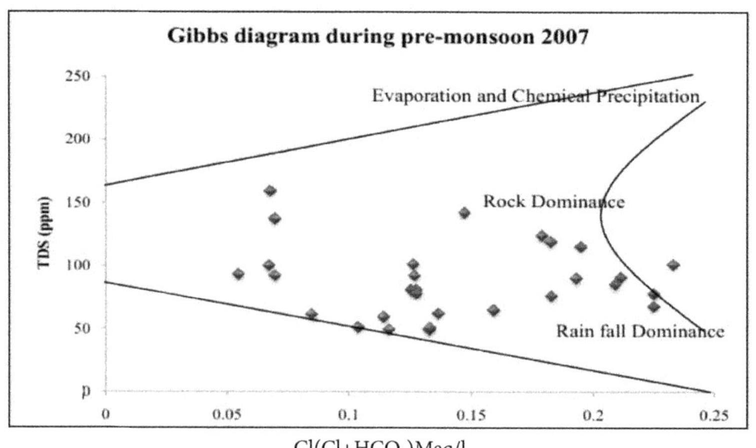

Cl(Cl+HCO_3)Meq/l

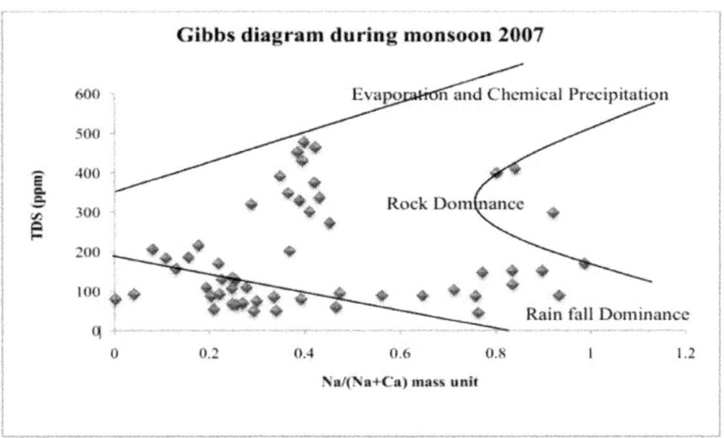

Fig. 4.13: Gibbs diagrams for the cations and anions of groundwater

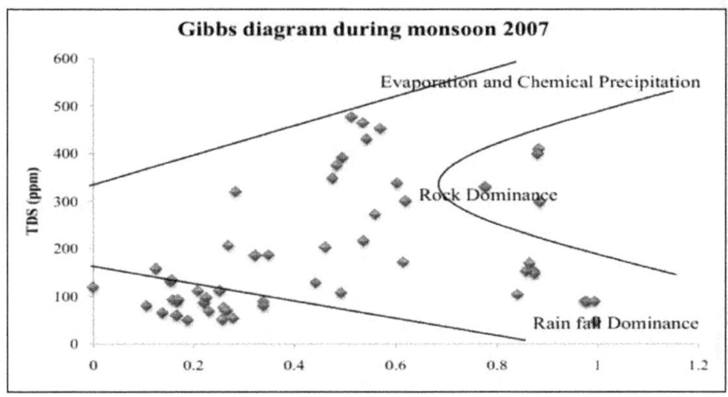

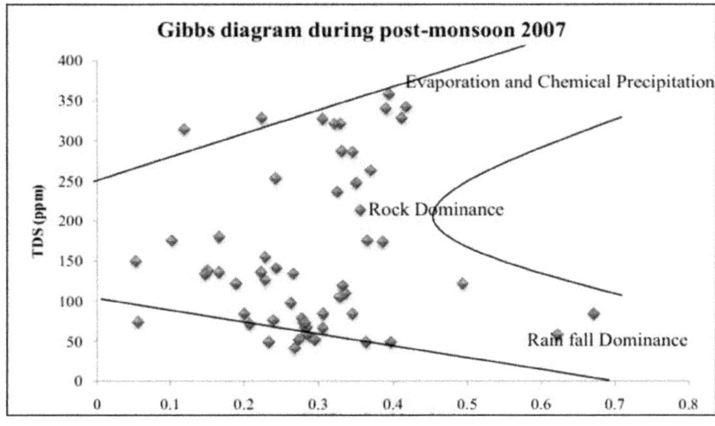

Cl(Cl+HCO$_3$)Meq/l

Fig. 4.14: Gibbs diagrams for the cations and anions of groundwater

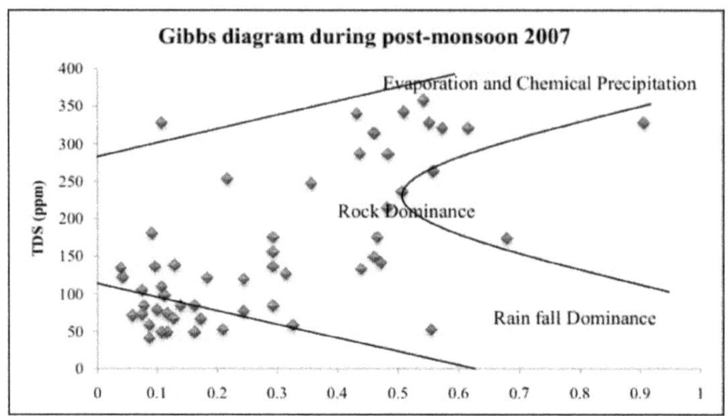

Cl(Cl+HCO$_3$)Meq/l

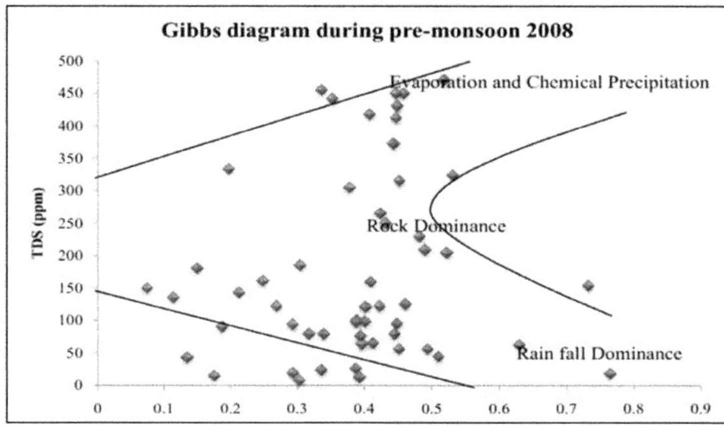

Na/(Na+Ca) Mass unit

Fig. 4.15: Gibbs diagrams for the cations and anions of groundwater

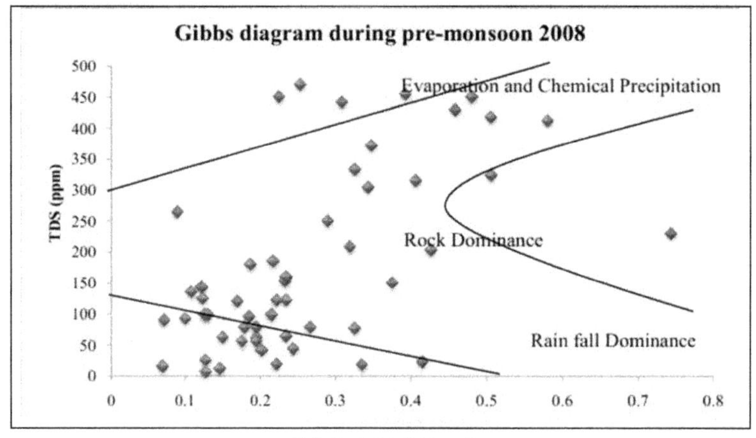

Cl/(Cl+HCO$_3$) Meq/l

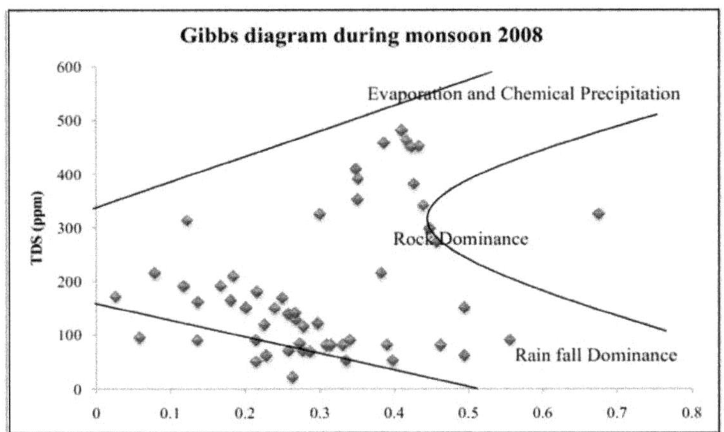

Na(Na+Ca) Mass unit

Fig. 4.16: Gibbs diagrams for the cations and anions of groundwater

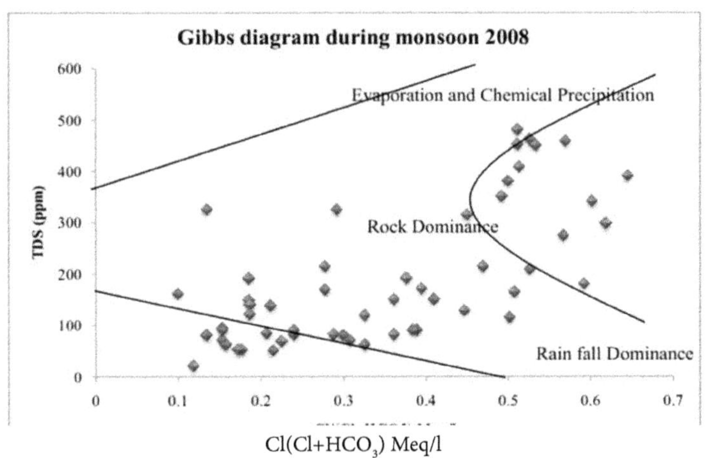

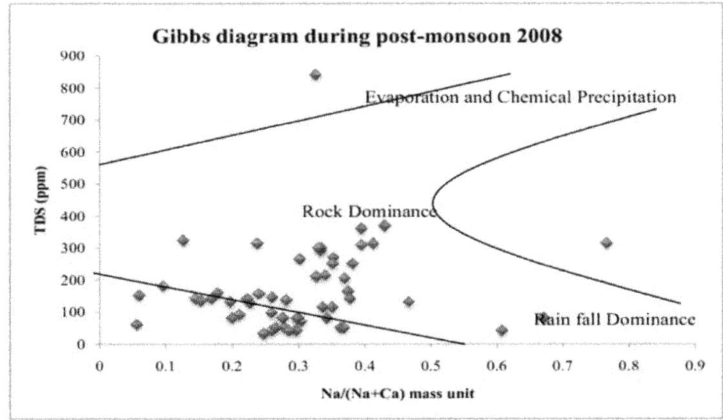

Fig. 4.17: Gibbs diagrams for the cations and anions of groundwater

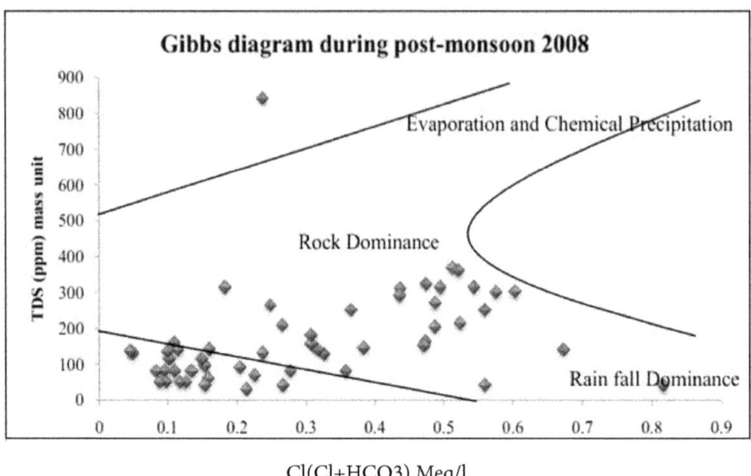

Cl(Cl+HCO3) Meq/l

Fig. 4.18: Gibbs diagrams for the cations and anions of groundwater

References

Zaporozec A, Graphical interpretation of water quality data. 10(2)ground water march-April, 1972.

Darnley AG, Björklund A, Bølvinken B, A Global Geochemical Database for Environment and Resource Management. Recommendations for International Geochemical Mapping. FinalReport of IGCP Project, 259, Earth Sciences, Paris, 1995.

Edmunds WM, Carillo-Rivera JJ, and Cardona A, Geochemical evolution of groundwater beneath Mexico City. *J Hydrol.* 258, 1– 24, 2002.

Babiker IS, Mohamed MAA, and Hiyama T, Assessing groundwater quality using GIS. Water Resour. Manage,21, 699– 715, 2007.

Marghade D, Malpe DB, Zade AB, Geochemical characterization of groundwater from northeastern part of Nagpur urban, Central India. Environ Earth Sci, 62, 1419–1430, DOI: 10.1007/s12665-010-0627-y, 2011.

Piper AM, A graphic procedure I the geo-chemical interpretation of water analysis. USGS Groundwater Note no,12, 1953.

Piper AM, A geographic procedure in the geochemical interpretation of water analysis [J]. Trans. Am. Geophysics Union. Washington D.C. 25, 914–928, 1994.

Walton WC, Groundwater Resources Evaluation. Mc Graw HillBook Co., New York, 1970.

Somay AM, Gemici U and Filiz S, Hydrogeochemical investigation of Kucuk Menderes River coastal wetland, Selcuk– Izmir, Turkey. *Environ Geol,*16, DOI: 10.1007/s00254-007-0972-7, 2007.

Ramkumar, T, Venkatramanan, S, Mary IA, Tamilselvi M and Ramesh G, Hydrogeochemical Quality of Groundwater in Vedaranniyam Town, Tamilnadu, India. *Research Journal of Earth Sciences,* **1(1)**, 28-34, 2009.

Srinivasamoorthy K,Nanthakumar C, Vasanthavigar M,Vijayaraghavan K, Rajivgandhi R, Chidambaram S, Anandhan P, Manivannan R and Vasudevan S, Groundwater quality assessment from a hard rock terrain, Salem district of Tamilnadu, India. *Arab J Geosciences 4,* 91–102, DOI 10.1007/s12517-009-0076-7, 2011.

Ravikumar P and Somashekar RK, Geochemistry of groundwater, Markandeya River Basin, Belgaum district, Karnataka State, India. *Chin. J. Geochem.* 30, 051–074, DOI: 10.1007/s11631-011-0486-6, 2011.

Ramkumar T, Venkatramanan S, Mary IA, Tamilselvi M and Ramesh G, Hydrogeochemical Quality of Groundwater in Vedaraniyam Town, Tamil Nadu, India. *Research Journal of Environmental and Earth Sciences,* 2(1),44-48, 2010.

Raju NJ, Shukla UK and Ram P, Hydro-geochemistry for the assessment of groundwater quality in Varanasi: a fast-urbanizing center in Uttar Pradesh, India. *Environ Monit Assess,*173, 279– 300 DOI: 10.1007/s10661-010-1387-6, 2011.

Mondal NC, Singh VP, Singh S and Singh VS, Hydrochemical characteristics of coastal aquifer from Tuticorin, Tamil Nadu, India. *Environ Monit Assess*175, 531–550, DOI: 10.1007/s10661-010-1549-6, 2011.

Fianko JR, Adomako D and Osae S, Ganyaglo S, Kortatsi BK, Tay CK and Glover ET, The hydrochemistry of groundwater in the Densu River Basin, Ghana. *Environ Monit Assess,*167, 663–674, DOI: 10.1007/s10661-009-1082-7, 2010.

Fianko ET, Osae S, Adomako D and Achel DG, Relationship between land use and groundwater quality in six districts in the eastern region of Ghana. *Environ Monit Assess,* 153, 139–146, DOI: 10.1007/s10661-008-0344-0, 2009.

Kumar SK, Rammohan V, Sahayam JD and Jeevanandam M, Assessment of groundwater quality and hydrogeochemistry of Manimuktha River basin, Tamil Nadu, India. *EnvironmentalMonitoring and Assessment,*159(1-4),341-351, DOI:10.1007/s10661-008-0633-7, 2008.

Jeevanandam M, Kannan R, Srinivasalu S and Rammohan V, Hydrogeochemistry and Groundwater Quality Assessment of Lower Part of the Ponnaiyar River Basin, Cuddalore District, South India. *Environmental Monitoring and Assessment,***132(1-3)**, 263-274, DOI: 10.1007/s10661-006-9532-y, 2007.

Subramani T, Rajmohan N and Elango L, Groundwater geochemistry and identification of hydrogeochemical processes in a hard rock region, Southern India. *Environmental Monitoring and Assessment,***162(1-4)**,123-137, DOI: 10.1007/s10661-009-0781-4, 2009.

Kumar M, Kumari K, Singh UK and Ramanathan AL, Hydrogeochemical processes in the groundwater environment of Muktsar, Punjab: conventional graphical and multivariate statistical approach. *Environmental Geology,***57(4)**, 873-884, DOI: 10.1007/s00254-008-1367-0, 2008.

Ravikumar P, Somashekar RK and Angami M, Hydrochemistry and evaluation of groundwater suitability for irrigation and drinking purposes in the Markandeya River basin, Belgaum District, Karnataka State, India. *Environmental Monitoring andAssessment*,**173(1-4)**,459-487, DOI: 10.1007/s10661-010-1399-2,2010.

Al-Ruwaih FM and Ben-Essa SA, Hydrogeological and Hydrochemical Study of the Al-Shagaya Field-F, Kuwait. *Bulletin of Engineering Geology and the Environment*, **63(1)**, 57-70, DOI:10.1007/s10064-003-0213-6, 2004.

Singh AK, Mondal GC, Kumar S, Singh TB, Tewary BK and Sinha A, Major ion chemistry, weathering processes and water quality assessment in upper catchment of Damodar River basin, India. *Environmental Geology***54(4)**, 745-758, DOI: 10.1007/s00254-007-0860-1.

Al-Ruwaih FM and Hadi KM, Hydrogeochemical study of the clastic aquifer in the Umm Gudair field, Kuwait. *Bulletin of Engineering Geology and the Environment*,**66(1)**,109-123, DOI:10.1007/s10064-006-0050-5.

Subba Rao N, Seasonal variation of groundwater quality in a part of Guntur District, Andhra Pradesh, India. *Environmental Geology*,**49(3)**,413-429, DOI: 10.1007/s00254-005-0089-9, 2006.

Gupta S, Mahato A, Roy P, Datta JK and Saha RN, Geochemistry of groundwater, Burdwan District, West Bengal, India. *Environmental Geology*, **53(6)**, 1271-1282, DOI: 10.1007/s00254-007-0725-7, 2008.

Manjusree TM, Joseph S and Thomas J, Hydro geochemistry and Groundwater Quality in the Coastal Sandy Clay Aquifers of Alappuzha District, Kerala. *J. Geo. Soc. of India*, **74**, 459-468, 2009.

Sami K, Recharge mechanism and Geochemical process in semi-arid sedimentary basins, Eastern Cape, South Africa. Jour.Hydrol., 27-48, 1992.

Subba Rao N, Geochemistry of groundwater in parts of Guntur district, Andhra Pradesh, India. Environ Geol,41, 552–562, DOI: 10.1007/s00254-003-0789-y, 2001.

Al-Agha MR and El-Nakhal HA, Hydrochemical facies of groundwater in the Gaza strip, Palestine. *Hydrologic Sci. J.*,**49(3)**, 359-371, 2004.

Bartarya SK, Hydrochemistry and rock weathering in a subtropical lesser Himalayan river basin in Kumaun, India. *J. Hydrology*,**146**, 149-174, DOI: 10.1016/0022-1694(93)90274, 1993.

Deutsch AW, Groundwater geochemistry: Fundamentals and applications to contamination. Lewis Publishers, New York, 1997.

Herlinger JR and Viero AP, Hydro-geochemistry of the Coxilha das Lombas Aquifer, Brazil. Environ. Chem. Lett.,5, 91-94, DOI: 10.1007/s10311-006-0083-9, 2007.

Kortatsi BK, A Hydro-chemical Framework of groundwater in the Ankobra Basin, Ghana. *Aquatic Geochemistry*, **13(1)**, 41-74, DOI: 10.1007/s10498-006-9006-4, 2006.

Marimuthu S, Reynolds DA and Le Gal La Salle C, A field study of hydraulic, geochemical and stable isotope relationships in a coastal wetlands system. *J Hydrol*,**315**, 93–116, 2005.

Ahmad Z and Qadir A, Source evaluation of physic-chemically contaminated groundwater of dera ismail khan area, Pakistan. *Environ Monit. Assess.*,**175**,**9**–21, DOI: 10.1007/s10661-010-1489-1, 2011.

Back W and Hanshaw BB, Advances in hydro-science. In Chemical Geohydrology [M]. Academic Press, New York, 11, 49,1965.

Rao GS and Rao GN, Study of ground water quality in greater Vishakhapatnam city, Andhra Pradesh (India). *J. Environ. Science & Engg*, 52(2), (a, b, c) 140, (d) 142, (e, f) 143, (g, h) 144, (i, j)145, 2010.

Verlecar XN, Desai SR, Sarcar A and Dalal, SG, Biological indicators in relation to coastal pollution along Karnataka coast, India. Water Res.,40, 3304-3312, 2006.

Murugesan A, Ramu A and Khanan N, Water quality assessment from Uthamapalayam Municipality in Theri district, Tamil Nadu, India. *Poll. Res.*,25, 163-166, 2006.

Chann CL, Zalifah MR and Norrakiah AS, Microbiological and physicochemical quality of drinking water. *The Malaysian Journal of Analytical Sciences,11*, 414-420, 2007.

Kumar VJ, Rao NA and pavanaguru R, Impact of pollution on the ground water in the developing scenario in north-eastern parts of Hyderabad city. *J. Env. Prot.,20*, 658-662, 2000.

Zutshi DP and Khan AV, Eutrophic gradient in Dal Lake, Kashmir. Indian *J. Env. Health*,30, 348-354, 1998.

Gibbs RJ, Mechanism controlling world water chemistry. *J. Science*. 170, 1088–1090, 1970.

Sami K, Recharge mechanism and geochemical process in semi-arid sedimentary basins, Eastern Cape, South Africa. *Jour.Hydrol.*, 27-48, 1992.

Meybeck M, Global Chemical weathering of surficial rocks estimated from river dissolved rock. *American Jour. Sci.*, 401-428, 1987.

5

Evaluation of Springs for Potability and Irrigation

This chapter is divided into Section-A and Section-B. The quality profile indices, chemical water quality index (CWQI) and heavy metal pollution index (HPI), of springs/hand pumps in terms of potability make the Section-A. The classification and distribution of different irrigational water quality parameters of the study area are covered in the Section-B. Salinity (EC), sodium adsorption ratio (SAR), soluble sodium percentage (SSP), residual sodium carbonate (RSC), magnesium hazards (MH) and permeability index (PI) are calculated for the evaluation of spring/hand pump water quality for irrigation. All samples are found within the permissible limit with respect to potability and irrigation except few samples throughout the study area.

Section A

5.1 Evaluation of springs/hand pumps for potability

5.1.1 Chemical Water Quality Index (*CWQI*)

The springs/hand pumps stand over-exploited so as to meet the increasing demands. Therefore, the evaluation of spring water quality is of great importance in determining the suitability of natural spring water for potability.

Monitoring programs of aquatic system plays a significant role in water quality control since it is necessary to know the degree of contamination in order to regulate its impact. Chemical water quality indices (CWQI) have been developed with the aim of providing summary information on quality. CWQI reflects the collective influence of various physico-chemical parameters and summarizes large amounts of water quality data into simple terms viz. excellent, good, fair and poor. The usefulness of CWQI is used as the indicator of water pollution in spatial and temporal comparisons of site(s). Literature survey reveals that a number of investigations have been conducted for studying the CWQI of several Indian rivers and ponds.

Chemical Water Quality Index (CWQI) relates water in terms of overall impact of various chemical parameters to its potability. Quality of water is defined in terms of its physical, chemical, and biological parameters. To study the trends of water quality in the study area, the CWQI with DO and without DO viz. $CWQI_{DO}$ and $CWQI_{WDO}$ of each sampling station is calculated. Quality indices are useful in obtaining a composite influence of the parameters of overall pollution. The CWQI is based on the sub index functions. Sub Index (SI) for each parameter is calculated by known mathematical functions, which are described in chapter 2. The parameters used are pH, TDS, total alkalinity, total hardness, dissolved oxygen (DO), Cl^-, Na, K, Ca, Mg, NO_3^-, and SO_4^{2-}. A comparative analysis revealed that the urban water quality was significantly poor as compared to the rural water.

For the comparison of urban and rural spring water, the seasonal variation of CWQI ($CWQI_{DO}$ and $CWQI_{WDO}$) is calculated table 5.1. Calculation of $CWQI_{DO}$ and $CWQI_{WDO}$ states enormous variation in rural and urban areas in the study area. This variation is further categorizes into different classes *viz.* excellent, good, fair and poor, and summarized in table 5.1. The water quality deterioration is high in urban compared to rural areas and it is due to high concentration of Na and K along with the dissolved oxygen (DO). DO have a very high impact on the indices, of the water quality throughout the study area. The high concentration of DO for springs of rural Almora reveals water of good quality than that of the urban Almora.

The percentage distribution of different water quality classes is illustrated in figure 5.1 and 5.2. Excellent class ranges from 39-40, 77-78 and 54-59% sample of springs with DO concentration and 61-70, 69-74 and 65-67% without DO concentration during pre-monsoon, monsoon and post-monsoon 2007-2008 respectively. This class of water quality is protected with a virtual threat or impairment and can be considered as extra clean fresh pristine water. Excellent class of spring water is best for human consumption without any kind of prior water treatment.

Table 5.1: Categorization of water quality and index range with

Class	Categorization	Index value	Remark
1.	Excellent	0-<45	Condition very close to extra clean fresh water resources; chemical water quality is out of impairment
2.	Good	46-<79	Chemical water quality deviates in minor degree from pristine level; minor purification is required before use
3.	Fair	80-<93	Chemical water quality is frequently threatened; consumption requires conventional treatment process before use
4.	Poor	Above >94	Unacceptable

Good class of water covers18-42 and 9-24 % of indices with DO and without DO respectively. Good class of water is protected with only a minor degree of threat whose condition rarely departs from desirable levels. The fair class of water indices ranges between 0-11% requires protection and such kind of condition departs from natural or desirable levels. Poor class of water quality is almost threatened or impaired which is unacceptable for potability. It has 0-13 % of total water indices which requires prior treatment before consumption. Apart from the effect of concentration of DO on water quality, Na^{+}, K^{+}, and NO_3^{-} also play an important role as water pollutants.

Table 5.2: Calculated CWQI with DO and without DO for urban (Almora town) as well as rurmonsoon, monsoon and post-monsoon 2007-2008.

	Sample ID Index		A1	A2	A3	A4	A5	A6	A7	A8
2007	With DO	Pre-Monsoon	135.81	126.03	94.02	100.13	97.48	85.54	81.44	74.15
		Monsoon	81.3	78.39	55.21	40.24	45.8	72.85	37.21	29.16
		Post Monsoon	102.4	96.11	52.39	58.36	56.14	51.29	50.75	55.62
	Without DO	Pre-Monsoon	169.35	154.83	84.21	88.31	94.33	116.06	66.12	39.96
		Monsoon	151.3	136.9	68.67	73.1	85.25	105.5	58.5	32.75
		Post-Monsoon	167.2	152	77.78	67.35	66.82	90.08	71.18	62.26
2008	With DO	Pre-monsoon	139.86	140.58	95.89	104.94	103.51	85.31	86.76	77.08
		Monsoon	96.88	77.78	41.27	44.58	55.93	74.22	48.38	30.92
		Post-Monsoon	98.41	92.34	49.94	45.93	58.39	53.56	43.57	48.19
	Without DO	Pre-Monsoon	184.75	177.94	88.54	92.67	99.19	120.4	68.14	42.02
		Monsoon	161.1	125.5	66.31	72.08	91.12	115.2	61.14	36.69
		Post Monsoon	175.9	148.4	69.29	70.81	76.17	89.6	75.51	66.12

Year		Sample ID Index	A9	A10	A11	A12	A13	A14	A15	A16
2007	With DO	Pre-Monsoon	81.43	54.69	109.08	91.53	68.91	67.69	86.20	92.51
		Monsoon	43.32	14.92	63.32	67.63	39.59	20.56	42.98	49.71
		Post-Monsoon	66.7	47.98	66.24	70.92	35.35	49.2	59.54	68.99
	Without DO	Pre-Monsoon	53.60	25.71	115.92	90.61	57.66	41.75	80.69	92.41
		Monsoon	47.56	20.15	107.1	80.46	50.36	32.89	65.63	83.54
		Post-Monsoon	72.11	33.68	110.7	83.55	49.65	68.26	73.13	87.14
2008	With DO	Pre-Monsoon	86.52	55.51	120.72	93.76	73.74	62.86	87.47	94.88
		Monsoon	41.36	18.69	68.43	62.99	46.46	41.42	50.1	63.71
		Post-Monsoon	62.19	41.8	64.27	68.46	32.53	61.42	58.91	66.9
	Without DO	Pre-Monsoon	57.37	30.81	127.77	95.39	59.49	35.76	79.81	96.82
		Monsoon	54.74	21.8	113.7	86.74	56.4	36.99	71.38	87.88
		Post-Monsoon	74.4	36.35	114.1	86.03	55.13	72.96	77.38	92.04

Year		Sample ID Index	A17	M1	M2	M3	M4	M5	M6	S1
2007	With DO	Pre-Monsoon	33.66	19.38	50.67	63.81	77.22	66.73	35.73	100.42
		Monsoon	27.2	21.13	10.5	30.75	35.91	28.28	15.11	57.73
		Post-Monsoon	43.84	34.03	31.05	28.29	57.43	64.05	25.62	43.68
	Without DO	Pre-Monsoon	13.67	29.52	14.57	46.27	63.97	22.47	14.19	107.14
		Monsoon	16.09	21.38	10.57	41.08	57.87	16.75	11.98	94.9
		Post-Monsoon	25.97	22.06	16.53	11.38	54.86	20.54	13.59	25.67
2008	With DO	Pre-Monsoon	65.07	23.31	49.87	65.17	77.5	65.15	30.78	100.2
		Monsoon	20.18	21.16	21.96	35.36	48.59	22.00	20.23	58.4
		Post-Monsoon	44.06	26.84	22.2	19.53	59.42	62.87	27.44	80.69
	Without DO	Pre-Monsoon	25.09	33.46	17.76	47.46	65.88	25.4	14.9	106.7
		Monsoon	15.32	26.02	10.33	42.49	60.02	17.71	14.42	95.25
		Post-Monsoon	31.76	26.55	17.96	12.59	60.17	21.92	13.39	99.94

Sample ID Index			S2	S3	S4	S5	S6	S7	S8	P1
2007	With DO	Pre-Monsoon	68.09	40.61	39.26	61.03	68.25	57.29	17.71	16.83
		Monsoon	27.62	17.37	21.5	26.89	30.8	20.51	6.84	9.39
		Post-Monsoon	22.06	41.51	24.72	23.24	43.88	17.06	37.81	14.34
	Without DO	Pre-Monsoon	43.31	17.81	18.89	41.11	39.91	23.91	4.43	13.74
		Monsoon	35.26	16.18	16.69	35.69	39.39	16.65	3.76	10.3
		Post-Monsoon	21.33	35.99	15.5	30.7	36.82	15.6	7.57	10.55
2008	With DO	Pre-Monsoon	65.72	37.1	35.65	54.63	67.05	53.26	19.53	10.9
		Monsoon	37.84	21.01	30.62	25.56	29.29	16.71	18.41	22.84
		Post-Monsoon	31.13	42.06	21.06	19.98	42.51	15.33	35.72	15.12
	Without DO	Pre-Monsoon	46.75	20.3	19.56	42.63	44.83	2375	5.93	12.9
		Monsoon	31.03	16.65	17.51	36.88	41.01	16.83	3.75	10.3
		Post-Monsoon	25.55	41.47	15.84	31.72	39.56	18.65	8.18	13.79

Sample ID Index			P2	P3	P4	P5	P6	P7	J1	J2
2007	With DO	Pre-Monsoon	43.08	33.55	45.76	38.51	40.05	43.44	48.15	40.47
		Monsoon	15.33	9.99	12.49	16.97	13.58	21.16	22.27	19.21
		Post-Monsoon	17.25	27.82	27.01	33.73	17.98	22.07	32.79	14.04
	Without DO	Pre-Monsoon	16.67	11.92	23.71	28.52	24.04	19.41	24.49	17.55
		Monsoon	14.17	9.62	19.66	28	19.89	17.85	19.92	15.13
		Post-Monsoon	19.56	10.51	19.77	25.10	17.31	17.76	23.35	13.57
2008	With DO	Pre-Monsoon	42.48	28.16	40.02	39.02	39.68	39.85	54.49	37.45
		Monsoon	21.67	15.25	25.62	27.17	18.24	18.24	23.28	15.64
		Post-Monsoon	21.78	30.03	23.29	25.91	17.68	17.68	46.6	30.15
	Without DO	Pre-Monsoon	21.9	13.4	27.00	31.8	21.8	21.8	26.1	19.9
		Monsoon	17.2	14.2	17.33	19.1	21.47	21.47	20.89	16.64
		Post-Monsoon	21.48	14.59	20.02	24.68	18.55	18.55	27.4	14.85

Sample ID Index			J3	J4	J5	J6	HP1	HP2	HP3	HP4
2007	With DO	Pre-Monsoon	37.40	38.01	50.07	13.86	51.23	50.23	16.43	65.38
		Monsoon	19.16	7.79	16.28	11.8	40.79	4.51	47.18	66.04
		Post-Monsoon	32.31	20.59	58.81	51.54	50.66	10.95	43.59	60.6
	Without DO	Pre-Monsoon	15.43	9.30	29.24	15.93	10.13	15.44	19.14	36.47
		Monsoon	14.13	7.32	10.56	9.41	9.53	12.93	17.84	40.39
		Post-Monsoon	15.27	15.00	21.54	25.95	17.14	15.00	21.92	53.57
2008	With DO	Pre-Monsoon	47.5	38.94	95.17	42.62	42.83	40.04	41.22	50.66
		Monsoon	26.4	11.45	18.77	17.27	35.75	47.56	47.9	68.73
		Post-Monsoon	21.02	29.87	24.93	60.64	47.17	44.61	22.5	57.31
	Without DO	Pre-Monsoon	20.1	13.4	143.4	20.66	31.41	17.71	19.57	22.17
		Monsoon	17.01	8.85	13.41	19.59	12.09	18.48	17.39	43.6
		Post-Monsoon	15.8	14.34	23.07	26.73	19.58	15.72	23.04	56.41

Sample ID Index			HP5	HP6	HP7	HP8	HP9	HP10
2007	With DO	Pre-Monsoon	61.44	65.06	39.01	46.89	61.86	11.15
		Monsoon	13.17	15.83	24.19	51.73	51.98	19.04
		Post-Monsoon	53.03	24.48	55.41	50.93	47.01	33.27
	Without DO	Pre-Monsoon	15.85	22.48	17.88	36.14	20.93	16.27
		Monsoon	17.34	20.49	16.31	38.86	21.38	15.71
		Post-Monsoon	21.55	29.42	25.97	49.93	31.87	20.65
2008	With DO	Pre-Monsoon	86.05	51.51	34.34	81.9	57.16	36.64
		Monsoon	16.71	24.04	26.87	60.53	63.43	22.18
		Post-Monsoon	39.39	50.82	38.25	65.02	43.32	74.22
	Without DO	Pre-Monsoon	58.5	18.96	28.15	50.48	32.58	19.43
		Monsoon	18.64	21.7	17.94	40.88	2.87	18.03
		Post-Monsoon	23.09	29.98	29.9	55.51	34.13	96.65

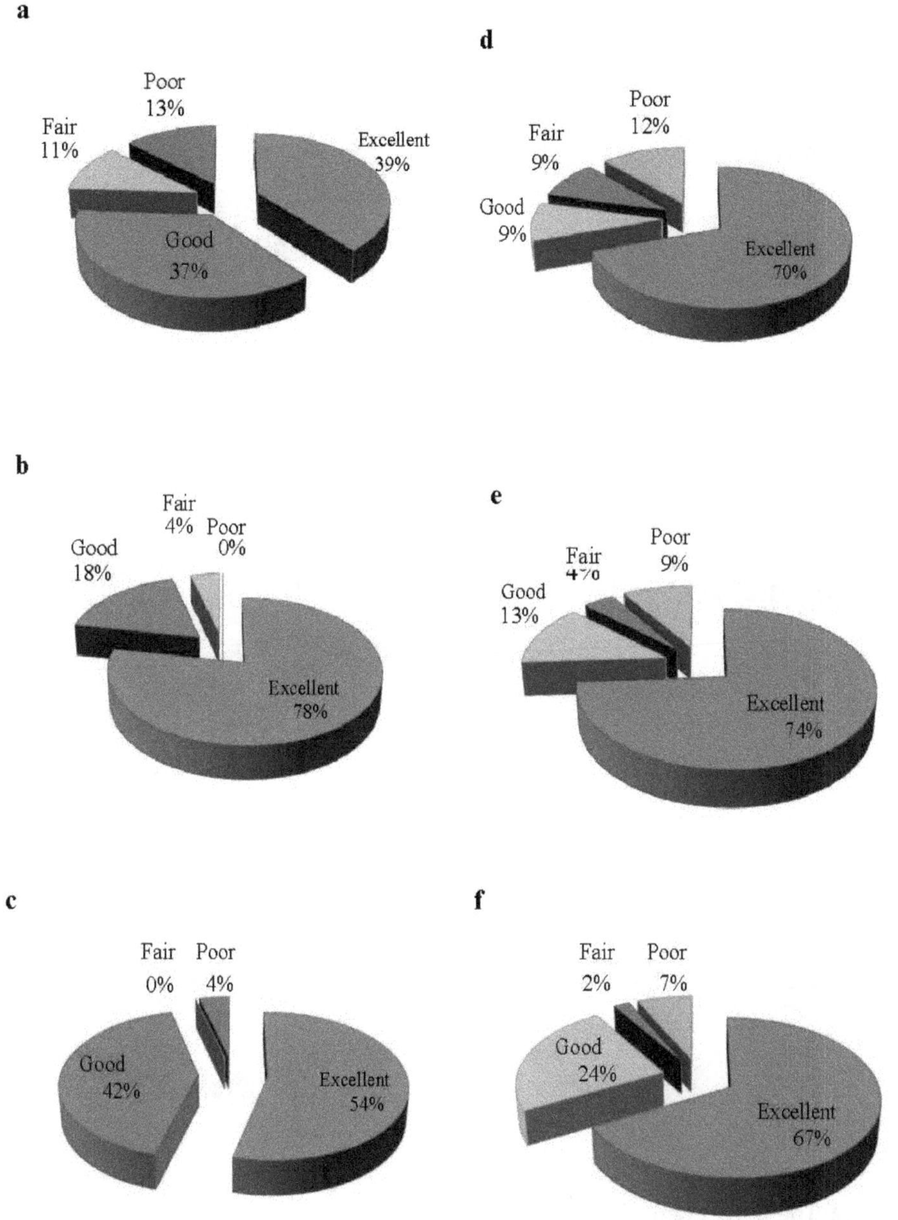

Fig. 5.1: The percentage distribution of $CWQI_{DO}$(with DO) and$CWQI_{WDO}$ (without DO) of different classes (a, d) during pre-monsoon2007 (b, e) during monsoon2007 (c, f) during post monsoon2007 respectively.

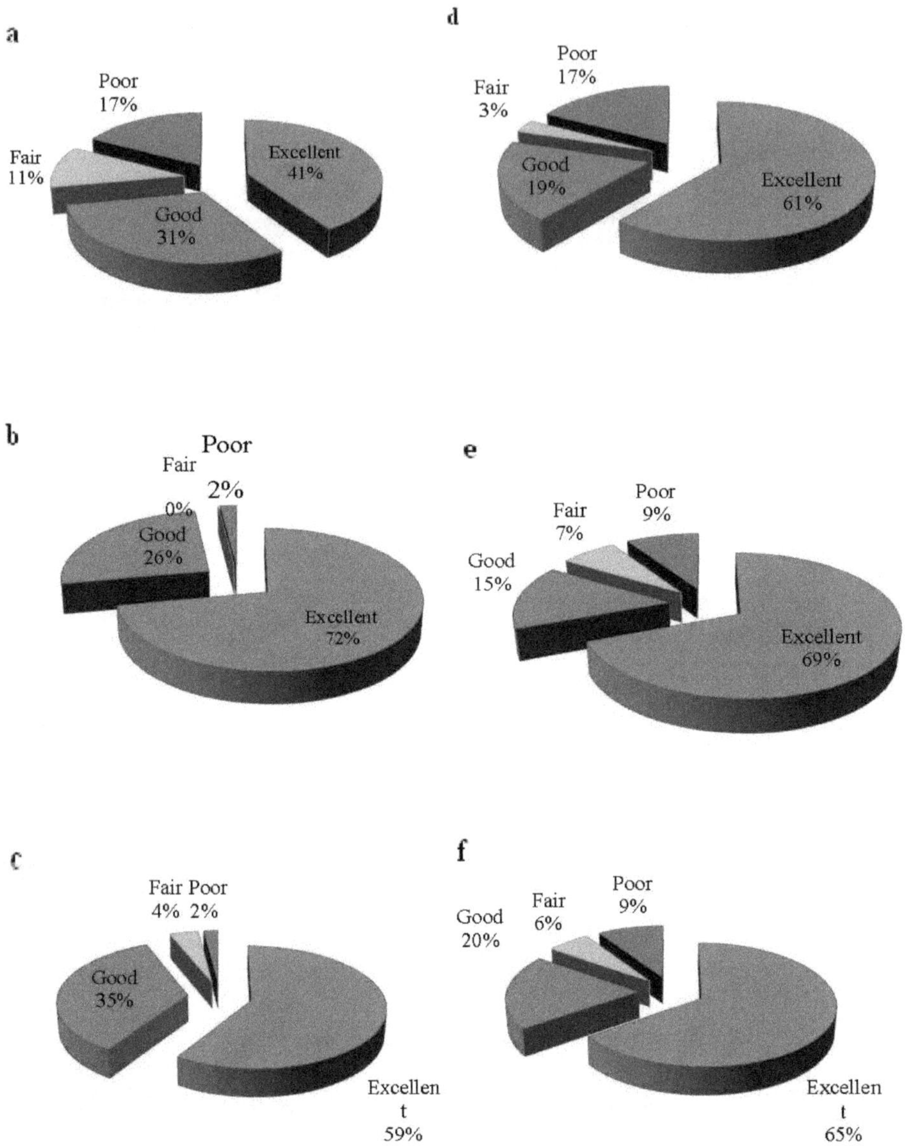

Fig. 5.2: The percentage distribution of $CWQI_{DO}$(with DO) and$CWQI_{WDO}$ (without DO) of different classes (a, d) during pre-monsoon 2008 (b, e) during monsoon 2008 (c, f) during post monsoon 2008 respectively.

Almora town spring water has higher degree of pollutant (Na^+, K^+, and NO_3^-) concentration. The high concentration of Na^+ and K^+ in Almora town spring water is attributed to its geological formation (Chapter-3). The reason for high nitrate in urban areas may be attributed to the percolation of fertilizers and improper handling of wastes.

Based on the Nitrate concentration, springs/Hand pumps of the study area have been classified as A, B and C (Figure 5.3). Class A contains nitrate concentration between 0-10 mg/l. Nitrate concentration fluctuates between 10-30 mg/l in class B and in class C nitrate is found in the range of 30-43 mg/l.

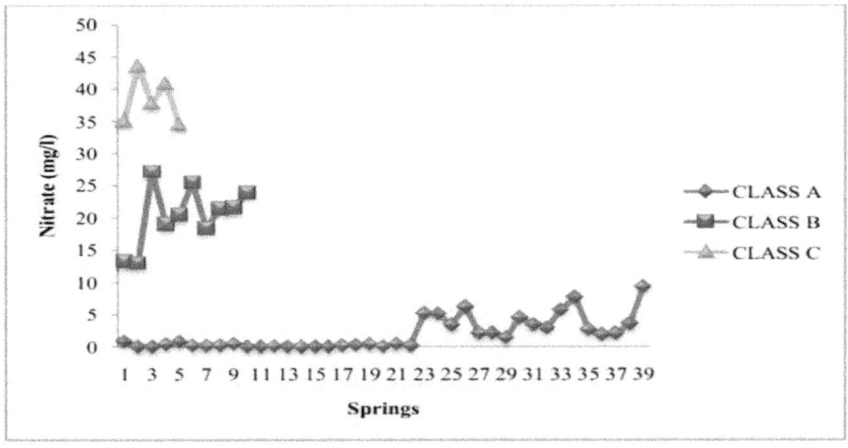

Fig. 5.3: Nitrate based groupings of springs

5.1.2 Heavy Metal Pollution Index (HPI)

Heavy Metal Pollution Index (HPI) plays an important role in quality indexing of spring/hand pump water as heavy metals are considered to be of potentially hazardous and cause physiological and neurological disorder. Literature survey reveals that investigations have been conducted to study the Heavy metal Pollution Index (HPI) of several Indian rivers and ponds.

Heavy Metal Pollution Index (HPI) is calculated by employing the weighted arithmetic average mean method of indexing. Heavy metals used for the HPI are Fe, Mn, Cu and Zn. HPI of the twenty springs of the study area with respect to the potability is calculated by the method described in the chapter 2.

Table 5.3 contains Heavy Metal Pollution Index (HPI) values for selected springs and hand pumps of the study area. The calculated index values indicate that spring water is not contaminated with respect to heavy metal ions. The critical pollution index value for drinkingwater is 100. Index values which are more than permissible limit 100 is not good for human consumption. The sample S-1 and M-1 crosses the limit (Table 5.3). High values of Zn (S-1= 7.66 mg/l; M-1= 17.065 mg/l) and Fe (S-1= 72.349 mg/l; M-1= 82.365mg/l) are recorded for S-1 and M-1 and this may be attributed to the poor condition of hand pumps.

Springs with index value more than 100 are not suitable for drinking with respect to heavy metals. The Heavy Metal Pollution Index (HPI) calculated for the spring water of the region has been found below the index limit of 100 except S-1 and M-1 and suggests that the spring water is not polluted with heavy metals.

Table 5.3: Heavy Metal Pollution Index (HPI) for selected samples.

Sample Name	HPI	Sample Name	HPI
A-7	27.041	A-8	68.769
A-11	81.476	A-10	17.359
S-4	6.375	S-1	**181.074**
S-6	33.342	S-5	36.578
S-8	11.208	S-7	40.004
P-1	36.926	**M-1**	**211.041**
P-3	29.037	P-2	29.037
J-1	28.652	P-7	32.008
M-2	43.857	J-4	33.154
M-4	37.556	M-3	36.457

The heavy metals concentration follows the order Fe > Mg > Zn > Cu. 67% of samples are found excellent (HPI < 40), 9% coincides well (HPI < 60), 15% are poor (HPI < 85) and 9% of sample needs prior treatment before consumption (HPI > 100; S-1, HPI = 181.074, M-1, HPI = 211.04). High HPI for spring S-1 and M-1 may be attribute to the poor condition of Hand pumps.

Section-B

5.2 Evaluation of springs/hand pumps for irrigation

Irrigation water quality is related to its effects on soils and crops and its management. High quality crops can be produced only by using high-quality irrigation water keeping other inputs optimal. Characteristics of irrigation water that define its quality vary with the source of the water. There are regional differences in water characteristics, based mainly on geology and climate.

The chemical constituents of irrigation water can affect plant growth directly through toxicity or deficiency, or indirectly by altering plant availability of nutrients. The water quality evaluation in the area of study is carried out to determine their suitability for irrigation. The suitability of groundwater for irrigation is contingent on the effects on the mineral constituents of the water on both the plant and the soil. In fact, salts can be highly harmful. They can limit growth of plants physically, by restricting the taking up of water through modification of osmotic processes. Also salts may damage plant growth chemically by the effects of toxic substances upon metabolic processes. Salinity, sodicity and toxicity generally need to be considered for evaluation of the suitability of groundwater for irrigation. Parameters such as electrical conductivity (EC), percent sodium (Na%), magnesium hazard (MH) and sodium adsorption ratio (SAR), Wilcox diagrams were used to assess the suitability of water for irrigation.

In addition to this combinations of the different water parameters were adopted worldwide (i.e. salinity (EC), SAR, SSP, RSC, MH and PI) for the evaluation of spring water quality for irrigation purpose and these are also discussed in this work.

5.2.1 Salinity (EC)

Excess salt increases the osmotic pressure of the soil water and produces conditions that keep off the roots from absorbing water. This results in a physiological drought condition. Even though the soil appears to have plenty of moisture, the plants may wilt because the roots do not absorb enough water to replace water lost from transpiration. For the purpose of diagnosis and classification, the total concentration of soluble salts (salinity hazard) in irrigation water is expressed in terms of electrical conductance (EC). Based on the EC, irrigation water has been classified into four categories (table 5.4).

Table 5.4: Classification of irrigation water based on salinity (EC)

Level	Salinity (EC) (µS/cm)	Hazard and limitations
C1	< 250	Low hazard; no detrimental effects on plants, and no soil build up expected.
C2	250 - 750	Sensitive plants may show stress; moderate leaching prevents salt accumulation in soil.
C3	750 - 2250	Salinity will adversely affect most plants; requires selection of salt-tolerant plants, careful irrigation, good drainage, and Leaching.
C4	> 2250	Generally unacceptable for irrigation, except for very salt tolerant plants, excellent drainage, frequent leaching and intensive management.

The distribution of spring water samples from the study area with respect to salinity is represented in table 5.8. Irrigation water classification has indicated that 86% of the water samples collected for three seasons during 2007 and 2008 belongs to the C1 level, an excellent category. No sample is found is termed in unsuitable category in terms of salinity for irrigation during the study period.

5.2.2 Sodium hazard/ Sodium Absorption Ratio (SAR)

The Na or alkali hazard is expressed in terms of sodium adsorption ratio. The main problem with high sodium concentration is its effect on soil permeability and water infiltration. Excess Na^+ in water produces the undesirable effects of changing soil properties and reducing soil permeability.

Sodium also contributes directly to the total salinity of the water and may be toxic to sensitive crops. The sodium hazard of irrigation water is estimated by the Sodium Absorption Ratio (SAR) and is calculated by the following formula:

$$SAR = \frac{Na^+}{\sqrt{\frac{Ca^{2+} + Mg^{2+}}{2}}} \qquad ...1$$

Continued use of water having a high SAR leads to a breakdown in the physical structure of the soil. The sodium replaces calcium and magnesium sorbed on clay minerals and causes dispersion of soil particles. Based on the Herman Bower classification (table 5.5).

Groundwater samples with SAR less than 10 have been considered suitable for irrigation to all type of soils. The SAR of ground water more than 10 causes permeability problems in shrinking and swelling of clay soils.

Table 5.5: Classification of irrigation water based on SAR values

Level	SAR	Hazard
S1	<10	No harmful effects from sodium.
S2	10-18	Appreciable sodium hazard in fine-textured soils of high CEC, but could be used on sandy soils with good permeability.
S3	18-26	Harmful effects could be anticipated in most soils and amendments such as gypsum would be necessary to exchange sodium ions.
S4	>26	Generally unsatisfactory for irrigation.

The distribution of groundwater samples from the study area with respect to SAR is represented in table 5.9. During pre-monsoon, monsoon and post-monsoon during 2007 and 2008, the SAR value of all the samples except few are found less than 10, and are classified as excellent for irrigation.

5.2.3 Soluble Sodium Percentage (% Na)

Sodium concentration is important in classifying irrigation water because sodium reacts with soil to reduce its permeability. When the concentration of Na^+ is high in irrigation water, Na^+ tends to be absorbed by clay particles, displacing Mg^{2+} and Ca^{2+} ions. This exchange process of Na^+ in water for Ca^{2+} and Mg^{2+} in soil reduces the permeability and eventually results in soil with poor internal drainage. Hence, air and water circulation is restricted during wet conditions and such soil are usually hard when dry. Soluble sodium percentage (SSP) is an estimation of the sodium hazard of irrigation water like SAR, but it expresses the percentage of sodium out of the total cations. Soluble sodium percentage (SSP) is calculated by the following formula.

The classification of the irrigation water according to Todd based on the soluble sodium Percentage is tabulated in table 5.6.

$$SSP= \frac{\left(Na^{+}+K^{+}\right)}{\left(Ca^{2+}+Mg^{2+}+Na^{+}+K^{+}\right)}\times 100$$

It is evident from the table 5.11 that about 51 samples of the study area are excellent to good for irrigation during pre-monsoon, monsoon and post monsoon 2007 and 2008.

Table 5.6: Classification of irrigation water based on soluble sodium percentage (SSP) values

Water class	SSP	EC (μS/cm)
Excellent	<20	>250
Good	20-40	250-750

Water class	SSP	EC (µS/cm)
Permissible	40-60	750-2000
Doubtful	60-80	2000-3000
Unsuitable	> 80	> 3000

5.2.4 Residual Sodium Carbonate (RSC)

The residual sodium carbonate (RSC) is a valuable parameter that has a great influence on the suitability of irrigation water. The RSC equals the sum of the bicarbonate and carbonate concentrations minus the sum of the calcium and magnesium ion concentrations, where the ions are expressed in meq/L.

$$RSC= (HCO_3^{-}+CO_3^{2-})-(Ca^{2+}+Mg^{2+}) \qquad ...3$$

As RSC increases, much of the calcium and some magnesium are precipitated from the solution when water is applied to soil. This results in increasing the sodium percentage and the rate of sorption of sodium on soil particles and increases the potential for a sodium hazard. The RSC based classification of the irrigation is presented in table 5.7 and similar characterization of the ground water of the study area is given in the table 5.12.

The RSC values clearly indicate that there is wide variation in salt concentration of the ground waters with respect to its irrigation water quality. Almost 2/3rd of the water resources studied is found with RSC values more than 2.5 and is unsuitable for irrigation.

Table 5.7: Classification of irrigation water based on RSC values

RSC	Hazard
<0	None.
0-1.25	Low, with some removal of calcium and magnesium from irrigation water.
1.25-2.50	Medium, with appreciable removal of calcium and Magnesium from irrigation water.
>2.50	High, with most calcium and magnesium removed leaving sodium to accumulate.

5.2.5 Magnesium Hazards (MH)

In natural waters, Mg in equilibrium state will adversely affect crop yields. The concept of magnesium hazard (MH) of irrigation water is proposed by Szabolcs and Darab and redefined by Raghunath. Mg Hazards calculated by the following formula.

$$MH= \frac{Mg^{2+}}{\left(Ca^{2+}+Mg^{2+}\right)}\times 100 \qquad ...4$$

Magnesium hazard value higher than 50meq/l is considered unsuitable for irrigation. All samples are found within the permissible limit <50 meq/l and suitable for irrigation purposes. Statistics of the spring water samples according to magnesium hazards (MH) is given in table 5.13.

5.2.6 Permeability Index (PI)

The permeability index (PI) is used to indicate the suitability of groundwater for irrigation. Doneen classified irrigation water based on the Permeability index. The classification divides water as class-I, II and III. The permeability index is calculated by the following formula and given in the table 5.10.

$$PI= \frac{Na^{+} + \sqrt{HCO_3^{-}}}{\left(Ca^{2+} + Mg^{2+} + Na^{+}\right)} \times 100 \qquad ...5$$

Class I and II water are categorized as good for irrigation with 75% or more of maximum permeability. Class III water is unsuitable with 25% of maximum permeability. In the present study, P.I. values vary from 24.58 to 189.57 during 2007 and from 36.4 to 207.5 during 2008. In terms of PI all sample are found suitable for irrigation.

The seasonal variation of average irrigational water quality parameters in the springs are depicted in figure 5.4 (during 2007) and 5.5 (during 2008).

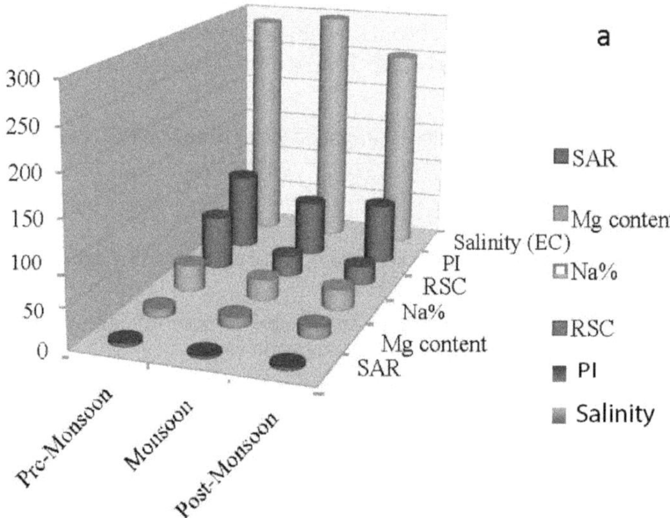

Fig. 5.4: Seasonal variation of average irrigational water quality parameters during 2007.

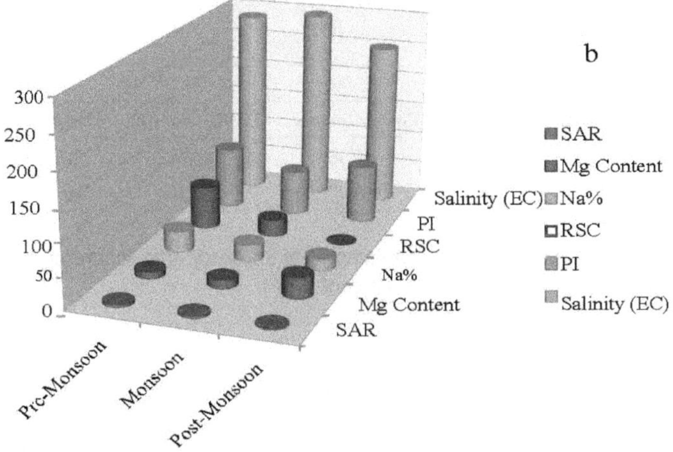

Fig 5.5: Seasonal variation of irrigational water quality during 2008.

Table. 5.8: Distribution of the spring water sample of the study area according to salinity (EC)

Level	Salinity (EC) (μS/cm)	2007			2008		
		Pre-Monsoon	Monsoon	Post-Monsoon	Pre-Monsoon	Monsoon	Post-Monsoon
C1	< 250	12-248 (35 samples)	74-247 (32 samples)	66-246 (34 samples)	71-232 (30 samples)	78-242 (29 samples)	71-241 (34 samples)
C2	250 - 750	252-721 (19 samples)	269-726 (22 samples)	272-563 (20 samples)	282-705 (23 samples)	258-732 (24 samples)	245-541 (20 samples)
C3	750 - 2250	-	-	-	762 (S1)	752 (A12)	-
C4	> 2250	-	-	-	-	-	-

Table 5.9: Distribution of the spring water samples of the study area according to Sodium Hazard (SAR)

Level	SAR	2007			2008		
		Pre-Monsoon	Monsoon	Post-Monsoon	Pre-Monsoon	Monsoon	Post-Monsoon
S1	<10	54 samples	53 samples	53 samples	52 samples	53 samples	53 samples
S2	10-18	-	1 sample (S1)	1 sample (S1)	1 sample (S1)	1 sample (S1)	1 sample (S1)
S3	18-26	-	-	-	1 sample (J5)	-	-
S4	> 26	-	-	-	-	-	-

Table 5.10: Distribution of the spring water samples of the study area according to Permeability Index (PI)

PI	2007			2008		
	Pre-Monsoon	Monsoon	Post-Monsoon	Pre-Monsoon	Monsoon	Post-Monsoon
Minimum	41.2	36.4	24.58	47.5	36.4	36.68
Maximum	158.7	186.6	189.57	207.5	191.1	185

Table 5.11: Distribution of the spring water samples of the study area according to Soluble Sodium Percent (SSP)

Sodium	Water class	2007			2008		
		Pre-Monsoon	Monsoon	Post-Monsoon	Pre-Monsoon	Monsoon	Post-Monsoon
<20	Excellent	3.99-19.35 (10 Samples)	0.02-18.97 (18 Samples)	4.25-19.98 (19 Samples)	4.27-18.96 (8 Samples)	1.96-9.8 (16 Samples)	2.74-19.79 (16 Samples)
40-60	Permissible	41.60-52.51 (4 samples)	40.52-52.44 (4 samples)	42.67-48.36 (4 samples)	40.18-59.26 (9 samples)	40.23-52. (3 sample)	47.84(S6)
60-80	Doubtful	-	60.28 (S1)	60.98 (S8)	67.57 (J5)	-	-
>80	Unsuitable	-	-	-	-	-	-

Table 5.12: Distribution of the spring water samples of the study area according to Residual Sodium Carbonate (RSC)

RSC (meq/l)	Remark on quality	2007			2008		
		Pre-Monsoon	Monsoon	Post-Monsoon	Pre-Monsoon	Monsoon	Post-Monsoon
<1.25	Good	-99.52-0.04 (11 samples)	-19.42-0.298 (6 samples)	-66.94-0.33 28 examples	-11.93-0.034 (J5, A16)	-15.37- -0.27 (5 samples)	-25--0.81 (9 samples)
1.25-2.50	Doubtful	-	2.39 (A1)	1.62 (A17)	1 sample (J5)	1.91 (A6)	1.69-2.18 (4 samples)
>2.50	Unsuitable	3.49-99.18 (43 samples)	4.72-116.00 (47 samples)	8.15-121.85 (25 samples)	6.74-138.91 (52 samples)	4.71-122.51 (48 samples)	2.96-123.65 (41 samples)

Table 5.13: Statisticsof the spring water samples of the study area according to magnesium Hazard (MH)

MH	2007			2008		
	Pre-Monsoon	Monsoon	Post-Monsoon	Pre-Monsoon	Monsoon	Post-Monsoon
Minimum	1.35	-5.15	-2.96	0.39	1.89	0.43
Maximum	46.00	25.31	40.28	27.45	33.33	38.37

5.2.7 Wilcox classification

Wilcox's diagram plots EC against Na% for classification of irrigation water. The observation based on Wilcox classification puts all the samples from excellent to good category. Wilcox classified spring water for irrigation purposes based on % Na and EC. All the Wilcox diagrams are plotted in figures 5.6 to 5.11. About 98% of the water resources are grouped as C1S1 (low-low) and C2S1 (medium-low) classes in pre-monsoon, monsoon and post monsoon of during 2007 and 2008. It is evident from all the Wilcox diagrams that Almora region samples are found in C2S1 (medium- low). This shows that the water resources other than Almora town region have low ionic concentration and have no salt effect.

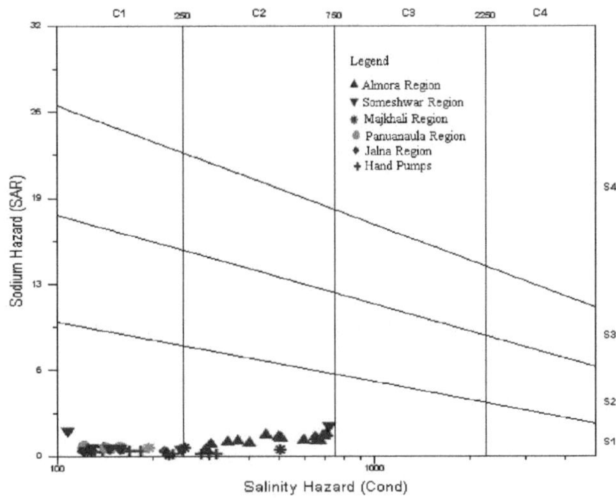

Fig. 5.6: Wilcox diagram of spring water during Pre monsoon2007

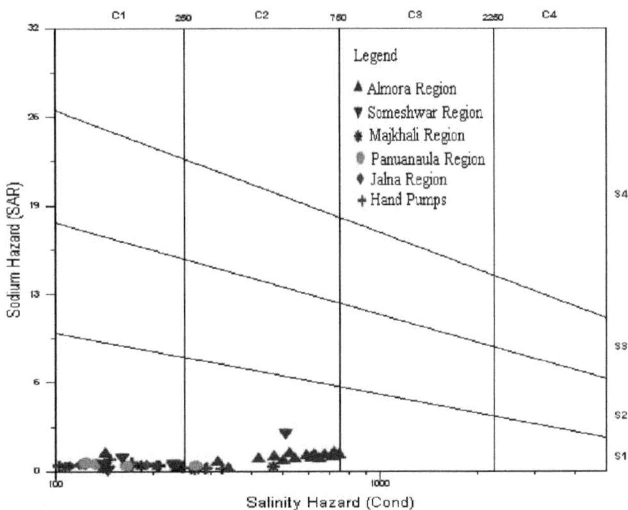

Fig. 5.7: Wilcox diagram of spring water during monsoon2007

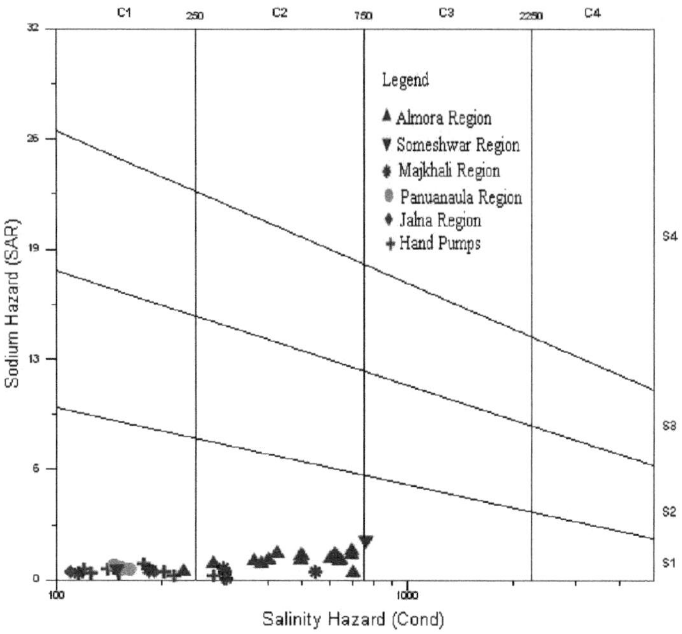

Fig. 5.8: Wilcox diagram of spring water during Postmonsoon2007

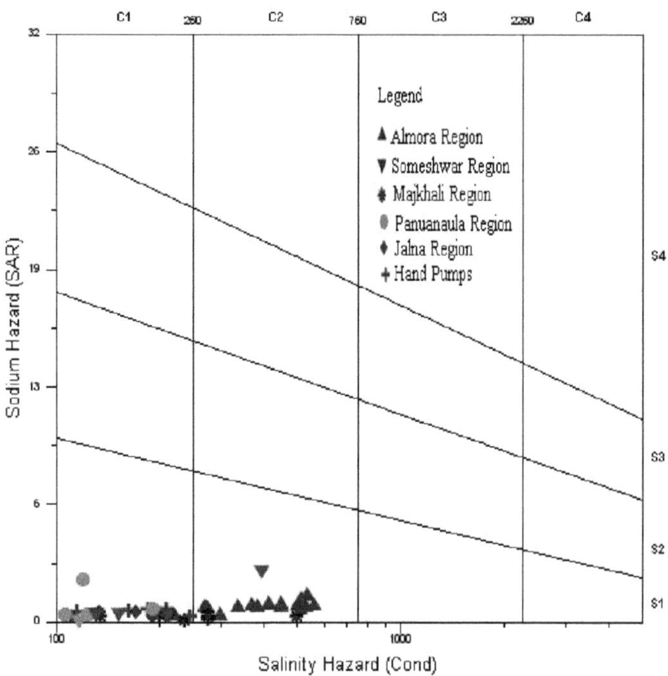

Fig. 5.9: Wilcox diagram of spring water during Pre monsoon2008

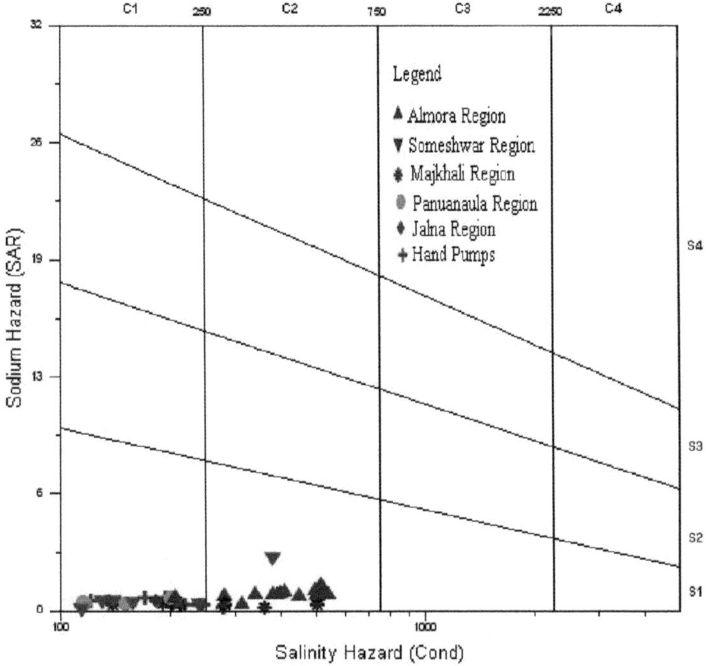

Fig. 5.10: Wilcox diagram of spring water during monsoon2008

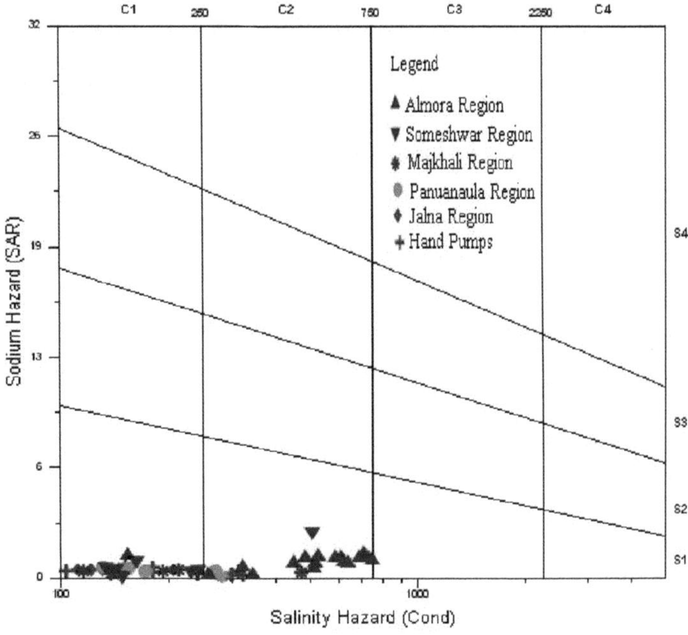

Fig. 5.11: Wilcox diagram of spring water during post monsoon2008

References

Almeida CA, Quintar S, Gonzalez P, and Mallea MA, Influence of urbanization and tourist activities on the water quality of the Potrero de los Funes River (San Luis—Argentina). Environmental Monitoring and Assessment, 133(1-3), 459–465,DOI: 10.1007/s10661-006-9600-3, 2007.

Boyacioglu H, Utilization of the water quality index method as a classification tool. Environmental Monitoring and Assessment, 167, 115-124, DOI: 10.1007/s10661-009-1035-1, 2010.

Sahu BK, Panda RB, Sinha BK and Nayak A, Water quality index of river Brahnnaniat Rourkela industrial complex of Orissa. J. Ecotoxico. Env. Monit,13, 169-175, 1991.

Khan F, Husain T and Lumb A, Water quality evaluation and trend analysis in selected watersheds of the Atlantic region of Canada. Environmental Earth Sciences,88(1-3), 221-248, DOI: 10.1023/A: 1025573108513, 2003.

Bordalo AA, Teixeira R and Wiebe WJ, A water quality index applied to an international shared river basin: the case of the Douro River. Environmental Management,38(6), 910-920, DOI: 10.1007/s00267-004-0037-6, 2006.

Kannel PR, Lee S, Lee YS, Kanel SR, and Khan SP, Application of water quality indices and dissolved oxygen as indicators for river water classification and urban impact assessment. Environmental Monitoring and Assessment, 132, 93–110, DOI:10.1007/s10661-006-9505-1, 2007.

Prakirake C, Chaiprasert P and Tripetchkul S, Development of specific water quality index for water supply in Thailand. Songklanakarin J. Sci. Technol., 31(1), 91-104, 2009.

Rosemond SD, Duro DC and Dube M, Comparative analysis of regional water quality in Canada using the water quality index. Environmental Monitoring and Assessment, 156(1-4),223-240, DOI: 10.1007/s10661-008-0480-6, 2009.

Yakubo BB, Yidana SM, Emmanuel N, Akabzaa T and Asiedu D, Analysis of ground quality using water quality index and conventional graphical methods: the Volta region Ghana. Environmental Earth Sciences, 59(4), 867-879, DOI:10.1007/s12665-009-0082-9, 2009.

Saeedi M, Abessi O, Sharifi F and Meraji H, Development of groundwater quality index. Environmental Earth Sciences, 163(1-4), 327-335, DOI: 10.1007/s10661-009-0837-5, 2010

Singh DF, Studies on the water quality Index of some major rivers of Pune Maharashtra, Proc. Acad. Env, Bio,1(1), 61-66, 1992.

Murthy R, Venkatamohan GV, Harischandrap S and Karthikeyan J, A preliminary study on water quality of river Tungabhara at kurnool town. Ind. J. Env. Prot.,14, 8, 1994.

Mariappan P, Vasudevan T and Yegnaraman V, Water quality in drinking water ponds (oorany) in eastern part of sivagongai district, Tamilnadu. J. Ind. wat. Works Assoc. April-june, 1999.

Chaturvedi MK and Bassin JK, Assessing the water quality index of water treatment plant and bore wells, in Delhi, India. *Environmental Earth Sciences,* **163(1-4),** 449-453, DOI: 10.1007/s10661-009-0848-2, 2010.

Majumdar D and Gupta N, Nitrate pollution of groundwater and associated human health disorders. *Indian J. Environ Hlth,* **42(1)**, 28-39, 2000.

Prasad B and Bose JM, Evaluation of the heavy metal pollution index for surface and spring water near a limestone mining area of the lower Himalayas. Environ Geo,41, 183-188, DOI: 10.1007/s002540100380, 2001.

Islam MS and Shamsad SZKM, Assessment of Irrigation Water Quality of Bogra District in Bangladesh. *Bangladesh J. Agril.Res.* 34(4), 597-608, 2009.

Ayres RS and Westcot DW. Water Quality for Agriculture. Irrigation and Drainage Paper No. 29. Food and Agriculture Organization of the United Nations. Rome. 1-117, 1985

Rowe DR and Abdel-Magid IM. Handbook of Wastewater Reclamation and Reuse. CRC Press, Inc., 550, 1995.

Khodapanah L, Sulaiman WNA and Khodapanah N, Groundwater Quality Assessment for Different Purposes in Eshtehard District, Tehran, Iran. European *Journal of Scientific Research,* **36(4)**,543-553, 2009.

Todd DK, Ground Water Hydrology. 2nd ed., John Wiley and Sons Inc. New York, USA,14,10-138, 1980.

Tank DK and Chandel CP S, Analysis of the Major Ion Constituents in Groundwater of Jaipur City. *Report and Opinion*,2(5), 1-7, 2010.

Shainberg I, Oster JD, —Quality of irrigation water‖, IIICpublication no. 2, 1976.

Subramani T, Elango L and Damodarasamy SR, Groundwater quality and its suitability for drinking and agricultural use in Chithar River Basin, Tamil Nadu, India, *Environ. Geol.*47, 1099–1110, 2005.

Karmegam U, Chidamabram S, Sasidhar P, Manivannan R, Manikandan S and Anandhan P, Geochemical Characterization of Groundwater's of Shallow Coastal Aquifer in and Around Kalpakkam, South India. *Research Journal of Environmentaland Earth Sciences,* **2(4)**, 170-177, 2010.

Al-Bassam AM, Al-Rumikhani YA, Integrated hydrochemical method of water quality assessment for irrigation in arid areas: application to the Jilh aquifer, Saudi Arabia. *Journal of AfricanEarth Sciences,* **36**: 345–356, 2003.

Sarkar AA and Hassan AA, water quality assessment of a ground water basin in Bangladesh for irrigation use. *Pakistan Journal of biological sciences,***9(9)**, 1677-1684, 2006.

Nishanthiny SC, Thushyanthy M, Barathithasan T and Saravanan S. Irrigation Water Quality Based on Hydro Chemical Analysis, Jaffna, Sri Lanka. American-*Eurasian J. Agric. & Environ. Sci.,* **7 (1),** 100-102, 2010.

Ali MS, Mahmood S, Chaudhary MN and Sadiq M, Irrigation quality of ground water of twenty villages in Lahore district. *Soil & Environ,* **28(1)**, 17-23, 2009.

Alexander JJ and Mahalingam B, Sustainable tank irrigation: An irrigation water quality perspective. *Indian Journal of Scienceand Technology*,**4(1)**, 22-26, 2011.

Almeida C, Quintar S, González P and Mallea M, Assessment of irrigation water quality: A proposal of a quality profile. *Environmental Monitoring and Assessment*, **142(1-3)**, 149-152,DOI: 10.1007/s10661-007-9916-7.

Vasanthavigar M, Srinivasamoorthy K, Vijayaragavan K, Ganthi RR, Chidambaram S, Anandhan P, Manivannan R and Vasudevan S, Application of water quality index for groundwater quality assessment: Thirumanimuttar sub-basin, Tamilnadu, India. *Environmental Monitoring and Assessment*, **171(1-4)**, 595-609, DOI: 10.1007/s10661-009-1302-1.

Vasanthavigar M, Srinivasamoorthy K, Ganthi RR, Vijayaraghavan K and Sarma VS, Characterization and quality assessment of groundwater with a special emphasis on irrigation utility: Thirumanimuttar sub-basin, Tamil Nadu, India. *Arabian Journal of Geosciences*, DOI: 10.1007/s12517-010-0190-6.

Jalali M, Hydro geochemistry of Groundwater and Its Suitability for Drinking and Agricultural Use in Nahavand, Western Iran. *Natural Resources Research*, **20(1)**, 65-73, DOI:10.1007/s11053-010-9131-z.

Doneen LD, Salinization of soils by salt in irrigation water. *Trans. Amer. Geophy. Union*, 35, 943-950, 1954.

Doneen LD, Quality of water for irrigation. In Proc. Conf. on Quality of Water for Irrigation. Water Resources Center,California, 14, 208, 1958.

Richards LA, Diagnosis and improvement of saline and alkali soils. US Department of Agriculture Handbook, 60, 1954.

Gholami S and Srinkantaswamy S, Analysis of Agricultural impact on the Cauvery River water around KRS Dam. *World applied Sciences Journal*, 6(8), 1157-1169, 2009.

Kelley WP, Alkali Soil–Their Formation Properties and Reclamation. Reinold Publication, New York, 124-128, 1946.

Herman Bower, International Student Edition, *GroundwaterHydrology*, 1978.

Salesh A, Al-Ruwaih F and Shehata M, Hydrogeochemical processes operating within the main aquifers of Kuwait. *J. Arid Environ*,42, pp.195-2091999.

Nagaraju A, Suresh S, Killham K and Hudson-Edwards K, Hydrogeochemistry of Waters of Mangampeta Barite Mining Area, Cuddapah Basin, Andhra Pradesh, India. *Turkish J. Eng.Env. Sci.*, 30, 203–219, 2006.

Collins R and Jenkins A, The impact of agricultural land use on stream chemistry in the middle Hills of Himalaya, *Nepal. Jour. Hydrol., 185(71),86*, 1996.

Bokhari AY and Khan MZA, Deterministic modeling of Al-Madinah Al-Munawarah ground water quality using lumped parameter approach. *Journal King Abdul Aziz University Earth Sciences*,5, 89-107, 1992.

Szabolcs I, Darab C, The influence of irrigation water of high sodium carbonate content of soils. In: proceedings of 8th International Congress of Isss, Trans II, 803-812, 1964.

Raghunath IIM, Groundwater. Second edition; Wiley Eastern Ltd., New Delhi, India, 344-369, 1987.

Doneen LD, Water quality requirement for agriculture. Proc. National Sym. Quality Standards for Natural Waters. Universityof Michigan ann. Report, 213-218, 1966.

Wilcox LV, Classification and use of irrigation water, U.S.O Department of Agriculture, Washington, 969, 1955.

6

Statistical Correlation Analysis of Springs

The significance of statistical methods in studying the hydrogeochemistry of the ground water is attempted in chapter six. Statistical correlation matrix for water quality parameters have been used to develop regression equations. The application of regression equations in predicting the concentration of highly related quality parameters completes the chapter.

Spring water is an important source of water supply throughout Kumaun region. A complete specification of water requires estimation of a large number of parameters. Moreover, for effective maintenance of water quality through appropriate control measures, continuous monitoring of quality parameter is essential. However, it is very difficult and laborious task for regular monitoring of all the parameters even if adequate manpower and laboratory facilities are available. Due to complexity of the chemical evolution of ground waters and sometimes substantially large amount of base information available, it is not easy to obtain a clear picture of the system under study.

As the routine chemical analysis requires lengthy and time consuming phenomena, it would be an attractive and significant solution to establish relationships between other different parameters with a common and easily determinable parameter. The developed regression equations for the parameters having significant correlation coefficients can be successfully used to estimate the concentration of others constituents.

More recently, some mathematical relationship between two individual parameters that are mutually and statistically correlated have been established. These correlations have been used for the prediction of parameters by regression equations. In addition to this, the linear correlation is very useful to get fairly accurate idea of quality of the ground water by determining a few parameters experimentally. In the present study a systematic calculation of correlation coefficient between each pair of water quality parameters was determined with regression analysis by determining correlation coefficient *(r)* with an aim to minimize the complexity of large set of data collected during the study period.

In the correlation regression study it is concluded that all the parameters are more or less correlated with each other.

6.1 Correlation Analysis

The collected parameters are projected in statistical analysis methods given in the chapter 2. Table 6.1, 6.2, 6.3, 6.4, 6.5, 6.6 contains the correlation coefficients matrix table analyzed during pre-monsoon, monsoon and post monsoon 2007-2008. Tables 6.1, 6.2, 6.3, 6.4, 6.5 and 6.6 represents high positive correlation between EC vs. TDS, TH, Na, Ca, Mg, Cl^-, and NO_3^-; between TDS vs. TH, Na, Ca, Mg, Cl^-, and NO_3^-; between TH vs. Ca, Mg, Cl^-, and NO_3^-; between Na vs. Ca, Cl^-, and NO_3^-; between Ca vs. Cl^-, NO_3^- and between TA vs. HCO_3^-. In the correlation analysis, it can be concluded that all the parameters are more or less correlated with each other.

One of the main objectives of the present investigation is to apply statistical correlation approach on physico-chemical parameters to find out regression equations to make theoretical calculation on water quality parameters of the study area. Correlation coefficient between X and Y (water quality parameters in the present case) was calculated for study area. If the numerical value of the correlation coefficient (r) between two variables X and Y is fairly large (>0.80), for these highly correlated parameters it is feasible to try a linear form as given below.

$$Y=AX+B \quad \ldots\ldots\ldots \quad 1$$

By knowing the values of A and B and one of the variable (for example X) one can predict the value of other variable (for example Y). Perfect correlation coefficient between different parameters based regression equations are given in tables 6.7, 6.8, 6.9, 6.10, 6.11 and 6.12. The linear correlation is very useful to get fairly accurate idea of quality of the springs/hand pumps by determining a few parameters experimentally. The observed and predicted values of selected samples are given in tables 6.13, 6.14, 6.15 and 6.16.

Table 6.1: Matrix table of the correlation coefficient during pre-monsoon 2007

Parameters	pH	EC	TDS	TH	Na	K	Ca	Mg	Cl^-	HCO_3^-	NO_3^-	SO_4^{2-}
pH	1	0.037	0.018	0.068	0.117	0.152	0.101	-0.07	0.006	0.257	-0.13	-0.12
EC		1	0.996	0.937	0.893	0.633	0.949	0.457	0.949	0.201	0.819	0.275
TDS			1	0.938	0.895	0.640	0.950	0.462	0.954	0.207	0.822	0.274
TH				1	0.788	0.549	0.978	0.617	0.883	0.244	0.699	0.285
Na					1	0.596	0.839	0.258	0.864	0.235	0.706	0.282
K						1	0.578	0.208	0.712	0.135	0.552	0.394
Ca							1	0.446	0.885	0.213	0.732	0.316
Mg								1	0.474	0.314	0.236	0.035
Cl^-									1	0.172	0.821	0.295
HCO_3^-										1	-0.13	0.207
NO_3^-											1	0.164
SO_4^{2-}												1

Table 6.2: Matrix table of the correlation coefficient during monsoon 2007

Parameters	pH	EC	TDS	TH	Na	K	Ca	Mg	Cl^-	HCO_3^-	NO_3^-	SO_4^{2-}
pH	1	0.247	0.247	0.251	0.065	0.189	0.264	0.206	0.207	0.138	0.103	-0.11
EC		1	0.999	0.952	0.567	0.182	0.846	0.829	0.936	0.389	0.857	0.210
TDS			1	0.952	0.565	0.182	0.847	0.829	0.937	0.389	0.855	0.211
TH				1	0.577	0.185	0.776	0.856	0.928	0.403	0.764	0.244
Na					1	0.588	0.133	0.316	0.466	-0.329	0.496	0.153
K						1	-0.04	0.033	0.075	-0.591	-0.03	0.172
Ca							1	0.679	0.767	0.649	0.693	0.176
Mg								1	0.859	0.442	0.710	0.233
Cl^-									1	0.376	0.816	0.215
HCO_3^-										1	0.256	0.093
NO_3^-											1	0.091
SO_4^{2-}												1

Table 6.3: Matrix table of the correlation coefficient during post-monsoon 2007

Parameters	pH	EC	TDS	TH	Na	K	Ca	Mg	Cl^-	HCO_3^-	NO_3^-	SO_4^{2-}
pH	1	0.053	-0.32	0.123	-0.04	-0.03	0.111	0.123	-0.07	0.329	-0.14	-0.03
EC		1	0.886	0.439	0.778	0.632	0.950	0.584	0.919	0.361	0.803	0.244
TDS			1	0.876	0.594	0.663	0.890	0.525	0.835	0.191	0.767	0.179
TH				1	0.673	0.615	0.973	0.621	0.897	0.379	0.726	0.300
Na					1	0.627	0.673	0.435	0.686	0.379	0.655	0.268
K						1	0.628	0.508	0.659	0.183	0.529	0.351
Ca							1	0.538	0.908	0.298	0.760	0.301
Mg								1	0.545	0.248	0.389	0.135
Cl^-									1	0.151	0.839	0.188
HCO_3^-										1	-0.03	0.099
NO_3^-											1	0.107
SO_4^{2-}												1

Table 6.4: Matrix table of the correlation coefficient during pre-monsoon 2008

Parameters	pH	EC	TDS	TH	Na	K	Ca	Mg	Cl^-	HCO_3^-	NO_3^-	SO_4^{2-}
pH	1	0.146	0.162	0.103	0.065	0.015	0.074	0.209	0.001	0.277	-0.14	-0.02
EC		1	0.981	0.855	0.710	0.567	0.848	0.616	0.785	0.415	0.691	0.276
TDS			1	0.819	0.667	0.542	0.817	0.569	0.743	0.427	0.686	0.293
TH				1	0.716	0.582	0.989	0.727	0.710	0.549	0.677	0.344
Na					1	0.534	0.726	0.447	0.574	0.259	0.522	0.264
K						1	0.589	0.364	0.628	0.258	0.446	0.646
Ca							1	0.620	0.702	0.524	0.670	0.350
Mg								1	0.545	0.501	0.507	0.201
Cl^-									1	0.408	0.408	0.408
HCO_3^-										1	0.198	0.259
NO_3^-											1	0.165
SO_4^{2-}												1

Table 6.5: Matrix table of the correlation coefficient during monsoon 2008

Parameters	pH	EC	TDS	TH	Na	K	Ca	Mg	Cl^-	HCO_3^-	NO_3^-	SO_4^{2-}
pH	1	0.240	0.230	0.228	0.285	0.119	0.201	0.212	0.211	0.176	0.129	-0.07
EC		1	0.989	0.951	0.828	0.619	0.921	0.825	0.934	0.564	0.861	0.268
TDS			1	0.952	0.828	0.616	0.926	0.815	0.941	0.555	0.858	0.242
TH				1	0.720	0.603	0.984	0.839	0.908	0.592	0.782	0.315
Na					1	0.538	0.672	0.656	0.727	0.555	0.691	0.289
K						1	0.565	0.483	0.621	0.259	0.517	0.457
Ca							1	0.803	0.903	0.579	0.751	0.309
Mg								1	0.720	0.695	0.638	0.207
Cl^-									1	0.348	0.868	0.222
HCO_3^-										1	0.188	0.241
NO_3^-											1	0.135
SO_4^{2-}												1

Table 6.6: Matrix table of the correlation coefficient during post-monsoon 2008

Parameters	pH	EC	TDS	TH	Na	K	Ca	Mg	Cl^-	HCO_3^-	NO_3^-	SO_4^{2-}
pH	1	0.067	-0.03	0.112	-0.08	0.19	0.097	0.103	-0.19	0.309	-0.16	-0.13
EC		1	0.543	0.952	0.538	0.018	0.948	0.737	0.710	0.439	0.777	0.200
TDS			1	0.555	0.293	0.008	0.555	0.425	0.374	0.105	0.469	0.091
TH				1	0.512	0.0261	0.976	0.827	0.719	0.461	0.735	0.210
Na					1	-0.003	0.525	0.359	0.455	0.258	0.549	-0.01
K						1	0.029	0.012	0.016	-0.047	0.048	0.065
Ca							1	0.686	0.714	0.408	0.757	0.191
Mg								1	0.562	0.484	0.512	0.214
Cl^-									1	0.119	0.697	0.464
HCO_3^-										1	0.039	0.059
NO_3^-											1	0.089
SO_4^{2-}												1

Table 6.7: Linear correlation coefficient and regression equation for some pairs of parasignificantly high value of coefficient during pre-monsoon 2007.

Pairs of parameters	*r*	Regression coefficient		Regression equations
		A	B	
EC vs.TDS	0.996	0.216	119.500	TDS=0.216×EC+119.5
EC vs.TH	0.937	0.083	56.260	TH=0.083×EC+56.26
EC vs. Na	0.893	0.024	9.657	Na=0.024×EC+9.657
EC vs.Ca	0.949	0.030	17.560	Ca=0.03×EC+17.56
EC vs. Cl^-	0.949	0.045	19.940	Cl=0.045×EC+19.94
EC vs.NO_3^-	0.819	0.017	3.674	NO_3^-=0.017×EC+3.674
TDS vs.TH	0.938	0.130	56.310	TH=0.13×TDS+56.31
TDS vs. Na	0.895	0.038	9.672	Na=0.038×TDS+9.672
TDS vs. Ca	0.950	0.046	17.590	Ca=0.046×TDS+17.59
TDS vs. Cl^-	0.954	0.070	19.930	Cl=0.07×TDS+19.93
TDS vs.NO_3^-	0.822	0.026	3.676	NO_3^-=0.026×TDS+3.676
TH vs. Cl^-	0.978	0.104	17.570	Cl=0.104×TH+17.57
TH vs. Ca	0.883	0.143	21.170	Ca=0.143×TH+21.17
Na vs. Ca	0.839	0.472	18.120	Ca=0.472×Na+18.12
Na vs. Cl^-	0.864	0.740	20.420	Cl=0.74×Na+20.42
Na vs. NO_3^-	0.864	0.264	4.103	NO_3^-=0.264×Na+4.103
Ca vs. Cl^-	0.885	0.455	20.820	Cl=0.455×Ca+20.82
Cl- vs. NO_3^-	0.821	0.153	3.459	NO_3^-=0.153×Cl+3.459

Table 6.8: Linear correlation coefficient and regression equation for some pairs of parameters which have significantly high value of coefficient during monsoon 2007.

Pairs of parameters	*r*	Regression coefficients		Regression equations
		A	B	
EC vs.TDS	0.999	0.205	125.1	TDS=0.205×EC+125.1
EC vs.TH	0.952	0.089	69.31	TH=0.089×EC+69.31
EC vs. Mg	0.829	0.005	2.866	Mg=0.005×EC+2.866
EC vs.Ca	0.829	0.025	19.05	Ca=0.025×EC+19.06
EC vs. Cl^-	0.857	0.016	3.577	Cl=0.016×EC+3.577
EC vs.NO_3^-	0.819	0.017	3.674	NO_3^-=0.017×EC+3.674
TDS vs.TH	0.952	0.14	69.32	TH=0.14×TDS+69.32
TDS vs. Mg	0.829	0.008	2.866	Mg=0.008×TDS+2.866
TDS vs. Ca	0.829	0.04	19.05	Ca=0.04×TDS+19.05

Pairs of parameters	*r*	Regression coefficients		Regression equations
		A	B	
TDS vs. Cl^-	0.937	0.067	23.04	Cl=0.067×TDS+23.04
TDS vs. NO_3^-	0.855	0.026	3.59	NO_3^-=0.026×TDS+3.59
TH vs. Cl^-	0.928	0.123	23.75	Cl=0.123×TH+23.75
Cl- vs. NO_3^-	0.816	0.133	3.611	NO_3^-=0.133×Cl+3.611
TH vs. Mg	0.856	0.016	2.895	Mg=0.016×TH+2.895
Mg vs. Cl^-	0.859	2.82	23.06	Cl=2.82×Mg+23.06

Table 6.9: Linear correlation coefficient and regression equation for some pairs of parameters which have significantly high value of coefficient during post-monsoon 2007.

Pairs of parameters	*r*	Regression coefficients		Regression equations
		A	B	
EC vs.TDS	0.886	0.156	121.6	TDS=0.156×EC+121.6
EC vs.Ca	0.950	0.032	25.31	Ca=0.032×EC+25.31
EC vs. Cl^-	0.919	0.044	17.16	Cl =0.044×EC+17.16
EC vs.NO_3^-	0.803	0.018	4.836	NO_3=0.018×EC+4.836
TDS vs.TH	0.876	0.144	81.46	TH=0.144×TDS+81.46
TDS vs. Ca	0.890	0.046	25.78	Ca=0.046×TDS+25.78
TDS vs. Cl^-	0.835	0.063	18.12	Cl=0.063×EC+18.12
TH vs. Ca	0.973	0.074	25.41	Ca=0.074×TH+25.41
TH vs. Cl^-	0.897	0.099	17.8	Cl=0.099×TH+17.8
Ca vs. Cl^-	0.908	0.315	17.72	Cl=0.315×Ca+17.72
Cl- vs. NO_3^-	0.839	0.189	4.044	NO_3^-=0.205×Cl+125.1

Table 6.10: Linear correlation coefficient and regression equation for some pairs of parameters which have significantly high value of coefficient during pre-monsoon 2008.

Pairs of parameters	*r*	Regression coefficients		Regression equations
		A	B	
EC vs.TDS	0.981	0.23	109.5	TDS=0.23×EC+109.5
EC vs.TH	0.855	0.075	59.25	TH=0.075×EC+59.25
EC vs.Ca	0.848	0.026	19.98	Ca=0.026×EC+19.98
TDS vs.TH	0.819	0.121	59.66	TH=0.121×TDS+59.66
TDS vs. Ca	0.817	0.043	20.08	Ca=0.043×TDS+20.08
TH vs. Ca	0.989	0.105	19.05	Ca=0.105×TH+19.05
TA vs. HCO_3^-	0.977	0.199	80.69	HCO_3^-=0.199×TA+80.69

Table 6.11: Linear correlation coefficient and regression equation for some pairs of parameters which have significantly high value of coefficient during post-monsoon 2008.

Pairs of parameters	r	Regression coefficients		Regression equations
		S	B	
EC vs.TH	0.952	0.099	83.18	TH=0.099×EC+83.18
EC vs.Ca	0.948	0.031	26.15	Ca=0.031×EC+26.15
TA vs. HCO_3^-	0.998	0.233	50.73	HCO_3^-=0.233×TA+50.73
TH vs. Ca	0.976	0.072	26.08	Ca=0.072×TH+26.08
TH vs. Mg	0.827	0.014	4.179	Mg=0.014×TH+4.179

Table 6.12: Linear correlation coefficient and regression equation for some pairs of parameters which have significantly high value of coefficient during monsoon 2008.

Pairs of parameters	r	Regression coefficients		Regression equations
		A	**B**	
EC vs. TDS	0.989	0.202	130.2	TDS=0.202×EC+130.2
EC vs. TH	0.951	0.089	72.15	TH=0.089×EC+72.15
EC vs. Na	0.828	0.02	9.964	Na=0.02×EC+9.964
EC vs. Ca	0.921	0.025	23.38	Ca=0.025×EC+23.38
EC vs. Mg	0.825	0.007	3.587	Mg=0.007×EC+3.587
EC vs. Cl^-	0.934	0.043	25.47	Cl=0.043×EC+25.47
EC vs. NO_3^-	0.861	0.017	3.803	NO_3^-3=0.017×EC+3.803
TDS vs. TH	0.952	0.141	71.83	TH=0.141×TDS+71.83
TDS vs. Na	0.828	0.031	9.902	Na=0.031×TDS+9.902
TDS vs. Ca	0.926	0.04	23.26	Ca=0.04×TDS+23.26
TDS vs. Cl^-	0.941	0.069	25.26	Cl=0.06×TDS+25.26
TDS vs. NO_3^-	0.858	0.026	3.769	NO_3^-=0.026×TDS+3.769
TA vs. HCO_3^-	0.999	0.192	52.57	HCO_3^-3=0.192×TA+52.57
TH vs. Ca	0.984	0.078	23.17	Ca=0.078×TH+23.17
TH vs. Mg	0.839	0.022	3.631	Mg=0.022×TH+3.631
TH vs. Cl^-	0.908	0.122	26.3	Cl=0.122×TH+26.3
Ca vs. Mg	0.803	0.064	3.793	Mg=0.064×Ca+3.793
Ca vs. Cl^-	0.903	0.376	26.79	Cl=0.376×Ca+26.79
Cl- vs. NO_3^-	0.868	0.137	3.515	NO_3^-=0.137×Cl+3.515

Table 6.13: The observed and predicted values of selected samples during pre-monsoon 2007

Sample ID	EC	TDS		TH		Ca^{2+}		Cl^-	
	Observed Value	Observed Value	Predicted	Observed Value	Predicted Value	Observed Value	Predicted Value	Observed Value	Predicted Value
A10	295	186	183.22	84	80.75	25.52	26.41	35.46	33.22
A14	346	220	194.24	76	84.98	25.12	27.94	35.46	35.51
A17	224	147	167.88	75	74.85	24.41	24.28	28.36	30.02
S5	248	181	173.07	78	76.84	23.41	25	35.46	31.1
H2	285	171	181.06	102	79.92	24.31	26.11	39	32.77
	TDS	TH		Na		Ca^{2+}		Cl	
A10	186	84	80.49	10.78	16.74	25.52	26.15	35.46	32.95
A14	220	76	84.91	23.24	18.03	25.12	27.72	35.46	35.33
S5	181	78	79.84	9.21	16.55	23.41	25.92	35.46	32.6

Table 6.14: The observed and predicted values of selected samples during monsoon 2007

Sample ID	EC	TDS		TH		Ca^{2+}		Mg^{2+}	
	Observed Value	Observed Value	Predicted	Observed Value	Predicted Value	Observed Value	Predicted Value		
A10	232	185	172.66	90	89.96	29.25	24.86	4.03	4.03
A14	369	230	200.75	78	102.15	26.08	28.28	5.49	4.71
J5	304	153	187.42	98	96.37	34.92	26.66	2.64	4.39
	TDS	TH		Mg^{2+}		Ca^{2+}		Cl^-	
A8	209	94	98.58	5	4.54	29.25	27.41	70.91	37.04
A13	324	96	114.368	5.49	5.46	29.25	32.01	42.46	44.75

Table 6.15; The observed and predicted values of selected samples during post-monsoon 2007

Sample ID	EC	TDS		TH		Ca^{2+}	
	Observed Value	Observed Value	Predicted	Observed Value	q	Observed Value	Predicted Value
A7	277	175	164.81	42.55	29.35	34.48	34.17
M3	235	235	158.26	35.46	27.5	37.85	32.83
	TDS	TH		Ca^{2+}		C^{l-}	
A7	175	114	106.66	34.48	33.83	42.55	29.15
M3	149	108	102.92	37.85	32.63	35.46	27.51

Table 6.16: The observed and predicted values of selected samples during pre-monsoon 2008

Sample ID	EC	TDS		TH		Ca^{2+}	
	Observed Value	Observed Value	Predicted	Observed Value	Predicted Value	Observed Value	Predicted Value
A7	385	385	198.05	80	88.13	29.25	29.99
A10	232	232	162.86	90	76.65	29.25	26.01
		TDS		TH		Ca^{2+}	
A7		250		80	89.91	29.25	38.93
A10		185		90	82.05	29.25	28.04

References

Kumar K, Rawat DS and Joshi R, Chemistry of springs in Almora, Central Himalaya, India. *Env. Geol*,**28(2)**, 1-7, 1996.

Usunoff EJ and Guzman-Guzman A, Multivariate analysis in hydrochemistry: An example of the use of factor and corresponding analysis. *Ground water*, **27(1)**, 27-34, 1989.

Joarder MAM, Raihan F, Alam JB and Hasanuzzaman S, Regression Analysis of Ground Water Quality Data of Sunamganj District, Bangladesh. *Int. J. Environ. Res.*,**2(3)**, 291-296, 2008.

Sharma CR, Chauhan P and Bahuguna M, Impact of Tehari dam on aquatic macro invertebrate diversity of Bhagirathi, Uttarakhand (India). *Journal of Environ, Science and Engg.* **50(1)**, 41-50, 2008.

Rajmohan NE, Ramachandran LS and Natarajan M, Major Ion correlation in ground water of Kancheepuram Region, South India. *Indian J. Environ Hlth*,**45(1)**, 5-10, 2003.

Pathak JK, Alam mohd, and Sharma S, Interpretation of ground water quality using multivariate statistical technique in Moradabad city, western Uttar Pradesh state, India. *E-Journal of Chemistry*, 5(3), 607-619, 2008.

Kowalkowski T, Kbyniewski R, Szpejna J and Buszewski B, *Water res.*,49, 74-75, 2006.

Padro REB, Castrillejo Y, Valasco MA and Vaga M, Anal. lett,26, 2617-2639, 1993.

Chenini I and Khemiri S, Evaluation of ground water quality using multiplelinear regression and structural equation modeling. *Int. J. Environ. Sci. Tech.*,**6 (3)**, 509-519, 2009.

Belkhiri L, Boudoukha A, Mouni L and Baouz T, Multivariate statistical characterization of groundwater quality in Ain Azel plain, Algeria. *African Journal of Environmental Science andTechnology*, **4(8)**, 526-534, 2010.

Reza R and Singh G, Physico-Chemical Analysis of Ground Water in Angul-Talcher Region of Orissa, India. *Journal ofAmerican Science*, **5(5)**, 53-58, 2009.

Ubale MB, Farooqi Mazahar, Arif Md. Pathan, Zahher Ahmad and Dhule DG, Regression analysis of ground water quantity data of chiklthana industrial area, Aurangabad (Maharashtra). *Oriental Journal of chemistry,* **17(2)**, 347-348, 2001.

Sarkar M, Banerjee A, Pramanick PP and Chakraborty S, Appraisal of elevated fluoride concentration in ground water using statistical correlation and regression study. *J. Indian Chem. Soc,* **83**,1023-1027, 2006.

Mishra PC, Pradhan KC and Patel RK, Quality of water for drinking and agriculture in and around mines in Keonjhar District, Orrisa. *Indian, J. Environ Hlth.*,**45(3)**, 273-220, 2003.

Jothivenkatachalam K, Nithya A and Chandra MS, Correlation analysis of drinking water quality in and around Perur block of the Coimbatore district, Tamil Nadu, India. *Rasayan J. Chem.*, **3(4)**, 649-654, 2010.

Dash JR, Dash PC and Pata HK, IJEP, 26(6), 550, 2006.

Mulla JG, Farooqui M and Zaheer A, *Int. J. Envi. Sci.*,**5(2)**, 943, 2007.

Kumar N and Sinha DK, *Int. J. of Envi Sci.***1(2)**, 253, 2010.

7

Summary

The seventh chapter entitled summary deals with a brief mention of the nature of the present work, its findings along with the scope for the further work in the field.

Water is one of the most precious gifts and first creation of Almighty. Water is the first and foremost important constituents of each and every living being. The dawn of civilization took place near the water resources. Water covers almost two third of the earth's surface. Water is without doubt the most abundant, most accessible, and the most studied of all chemical compounds. Its omnipresence, it's crucially important for man's survival and its ability to transform so readily from the liquid to the solid and gaseous states has ensured its prominence in man's thinking from the earliest times. In a sense, water is a weird compound of nature.

The first chapter entitled introduction, a review of the literature on various aspects of chemical composition of ground water, hydro-geochemical characterization, potability, irrigation and statistical studies of water quality.

In the chapter two various materials and methods employed in the present study are discussed. For collection, storage, preservation, sampling, and analysis of samples the method given in APHA and BIS were adopted. Temperature, dissolved oxygen, pH, EC of spring-water samples was measured on site by water potable analysis kit (Model MAC 12831). Alkalinity, hardness, chloride, bicarbonate, calcium and magnesium were analyzed by standard titrimetric methods. Sodium and potassium were analyzed by flame photometer (Model Systronics 128), Nitrate and sulfate by spectrophotometrically (Model Systronics 106).

The study area of selected sampling stations is covered in the chapter three. Almora district lies between latitudes 29 37′N and 29.62 N and longitudes 79 40′E and 79.67 E. It has an average elevation of 1,651 meters (5,417 feet) and is located on a ridge at the southern edge of the Kumaun Hills of the Himalaya range. Geologically the district is located on the Almora nappe I, extended up to north and south thrust.

The study area comprises Someshwar, Majkhali, Panuanaula, Jalna and specially Almora town (urban) regions. There are total fifty four samples including ten hand pumps have been selected for the present investigation.

The water samples analyzed for various physico-chemical parameters are as follows:

- Temperature, pH, oxidation-reduction potential (ORP), conductivity (EC), dissolved oxygen (DO), total dissolved salt (TDS), free CO_2, Total hardness (TH), salinity, H_2S (ND)
- Anions and cations-sodium (Na^+), potassium (K^+), calcium (Ca^+), magnesium (Mg^+), nitrate (NO_3^-), chloride (Cl^-), sulfate (SO_4^{2-}), bicarbonate (HCO_3^-)
- Heavy metals- Fe^{2+}, Cu^{2+}, $Zn^{2+}$and Mn^{2+}

The hydrogeochemical characterization and classification of hill springs are the part of chapter four. The diagnostic chemical properties of ground water are presented by various methods. The most common of which are Piper tri linear diagram and Durov diagram. These diagrams are useful in screening and sorting large number of chemical data, major cations and anions such as Na^+, K^+, Ca^{2+}, Mg^{2+}, HCO_3^-, SO_4^{2-} and Cl^-, which makes interpretation easier. The main geological and hydrological factors which generally affect the ground water geochemistry include rainfall, climate, soil, air, aquifer lithology, saline water and flow pattern. Piper tri linear diagram and Durov diagram is a plot of seven cations and anions. Figures demonstrate that the water samples occupy a wide area of the Piper-diagram ranging from fresh to saline waters. 75% samples fall in the Ca^+, HCO_3^- region and remaining 25% samples fall in no dominance area. In Pre-monsoon two major facies are dominant in the region, these are: (I) Ca-Na-Cl-HCO_3 (II) Ca-HCO_3-Cl. monsoon 2007 and Post-monsoon 2007 demonstrate that majority of samples fall in the Ca^{++} and HCO_3^- region. For spring waters in Pre-monsoon 2008 the major water types found are Ca-Na-Cl-HCO_3, Ca-Na-Cl and Na-Ca-HCO_3 water types predominate in the region. However monsoon 2008 cations-anions composition exhibits pattern similar to monsoon 2007. Piper-tri linear diagrams for the spring/ hand pumps clearly supports that they fall in fresh water domain and these findings are substantiated by Durov diagrams.

Gibb's classification depicts the variation of Cl/Cl+HCO_3 and Na/ Na^+ Ca as a function of TDS. Which explains that low TDS and high Cl/Cl+HCO_3 water is controlled by rainfall. Intermediate TDS and low Cl/Cl+HCO_3 water is controlled by rock weathering. High TDS and high Cl/Cl+HCO_3 water is controlled by rock weathering. These correlations are indicative of the fact that the chemical nature of the water is controlled by the rock weathering.

The hill springs have maximum value of 477 mg/l which indicates that the water is fresh in nature. Similarly the hand pumps recorded maximum value 329 mg/l for TDS and safe for common water uses.

Evaluation of springs for potability and irrigation is covered in chapter five. Heavy Metal Pollution Index (HPI) is also calculated by employing the weighted arithmetic average mean method of indexing. Heavy metals used for the HPI are Fe, Mn, Cu and Zn. The concentration of heavy metals found well within the safe limit with respect to their health hazards.

The suitability of ground water for irrigation purposes depends upon its mineral constitutes. The general criteria for judging the quality are (i) total salt concentration as measured by electrical conductivity (EC) (ii) relative proportion of sodium to other principle cations as expressed by SAR (iii) Residual sodium carbonate (RSC) (iv) Na % and (v) permeability index (PI).

Wilcox classified ground water for irrigation purposes based on % Na and EC. All the Wilcox diagrams are plotted in Figure 14-19. About 98% of the water resources are grouped as C1S1 (low-low) and C2S1 (medium-low) classes in pre-monsoon, monsoon and post monsoon of during 2007 and 2008. It is visible from all the Wilcox diagrams that Almora region samples are found in C2S1 (medium-low). This shows that the water resources other than Almora region have low ionic concentration and have no salt effect.

The statistical correlation analysis of springs attempted given in chapter six. A complete specification of water requires estimation of a large number of parameters. Moreover, for effective maintenance of water quality through appropriate control measures, continuous monitoring of quality parameter is essential. However, it is difficult and laborious task for regular monitoring of all the parameters even if adequate manpower and laboratory facilities are available. More recently, some mathematical relationship between two individual parameters that are mutually and statistically correlated have been established. Collected parameters were projected in statistical relations. The relation of water quality parameters on each other in samples of water analyzed was determined with regression analysis by determining co-relation co-efficient *(r).*

Positive correlation between Ca-Na, Ca-K, Ca-Cl, and Cl-K is observed, and represents positive correlation between Na-Cl, Ca-Cl, and Cl-K is also observed. Ca has perfect positive correlation with Cl, Na, K.

One of the main objectives of the present investigation is to apply statistical correlation approach on physico-chemical parameters to find out regression equations to make theoretical calculation on water quality parameters of the study area. Correlation coefficient between X and Y (water quality parameters in the present case) was calculated separately for Almora town. If the numerical value of the correlation coefficient (r) between two variables X and Y is fairly large (>0.90), for these highly correlated parameters it is feasible to try a linear form as given below.

$$Y = AX + B$$

By knowing the values of A and B and one of the variable (for example X) one can predict the value of other variable (for example Y).

Perfect correlation exists in between conductivity-TDS, Conductivity-HRD, Conductivity-Cl, Conductivity-Ca, TDS-HRD, TDS-DO and TDS-Ca. for these correlations separate regression equations have been developed.

Conclusions and Further Scope of Work

The springs/hand pumps studied in the present study have the following findings:

The water temperature oscillates throughout the study was -6.4 to 22.8 and -7.1 to 20.2°C during 2007 and 2008 respectively.

pH recorded was from 5.2 to 8.12 and from 5.3 and 7.4 during 2007 and 2008 respectively. Almost more than 60% samples were acidic in nature.

The indicator of total ions (cations and anions) *i.e.* TDS was found between 41 and 840 and between 12 to 841 mg/l during 2007 and 2008 in ground water resources of the study area.

During the study period the average DO observed in the range from 3.26 to 13.8 mg/l.

Average Electrical conductivity (EC) measured during the study period was from 63 to 762µS/cm.

The concentration of major ions (cations and anions) usually follow the trend $HCO_3^- > Cl^- > Ca^{2+} > Na^+ > SO_4^{2-} > NO_3^- > Mg^{2+} > K^+$ for the ground water resources.

The concentration of NO_3^- was found higher in some of the springs than its permissible limit especially in Almora town region.

The water characteristic of springs was also evaluated with the help of Piper trilinear diagram and Durov diagram and the hydro-chemical facies of the spring of the study area are characterized by Ca-Na-Cl-HCO_3, Ca-Na-HCO_3-Cl and Ca-HCO_3-Cl water type.

According to Piper diagram all samples fall in the fresh water region and according to Gibbs diagram total cations and anions are rock dominance.

Chemical water quality with respect to selected chemical parameters was found suitable for drinking but higher concentration of nitrate in some of the springs deteriorates water potability. Ground waters of the area are not contaminated with heavy metal ions.

Spring water quality for irrigation purpose was found suitable. According to Wilcox diagrams about 98% of the samples were grouped within C1S1 (low-low) and C2S1 (medium-low).

Collected parameters were projected in statistical relations. The relation of water quality parameters on each other in samples of water analyzed was determined with regression analysis by determining co-relation co-efficient (r).

Positive correlation between Ca-Na, Ca-K, Ca-Cl, Cl-K, Na-Cl, Ca-Cl, and Cl-K was demonstrated by correlation matrix table. It is also found that Ca has perfect positive correlation with Cl, Na and K.

This study can be used in prediction of unknown chemical parameters with precision for other similar water resources of the region for which the linear regression equation is applicable.

The present work is area specific and conducted in Almora Tehsil of district Almora of Kumaun Hills. This work can be extended to springs of Uttarakhand hills to develop geo-chemical mapping of springs.

www.ingramcontent.com/pod-product-compliance
Ingram Content Group UK Ltd.
Pitfield, Milton Keynes, MK11 3LW, UK
UKHW021952270726
14060UKWH00002B/480

9 789387 057654